Advanced Electrical Installations

C. Shelton

Advanced Electrical Installations

LONGMAN

Addison Wesley Longman Limited
Edinburgh Gate , Harlow
Essex CM20 2JE, England
and Associated Companies throughout the world

© Addison Wesley Longman Limited 1996

First published 1996

British Library Cataloguing in Publication Data
A catalogue entry for this title is available from the British Library

ISBN 0-582-24618-0

Set by 4 in 10/12pt Times
Produced by Longman Singapore Publishers (Pte) Ltd
Printed in Singapore

To my wife, Shirley, with thanks

Contents

Preface

Advanced Electrical Installations has been written as a logical companion to the 2nd edition of *Electrical Installations* and embraces the revised *City and Guilds Electrical Installation Scheme 2360* while closely meeting the requirements of the new *National Vocational Standards* in Electrical Installation Engineering and the *16th Edition of the IEE Wiring Regulations*.

Throughout this book, details and data have been presented in a practical and straightforward fashion using an everyday language code which is both readable and easy to follow. Historical and background material, mixed with important facts and advice at a practical level, have been interwoven within the text providing additional interest to the reader. At the end of each chapter there is a review section together with a variety of self-assessment questions enabling examination and evaluation of the student's new-found knowledge relating to aspects and topics featured within the chapter. The contents of this book are of a practical and theoretical nature. However, the mathematics used have been purposely kept to a minimum and used only when interpreting or explaining electrical concepts within the framework of electrical installation and engineering.

Advanced Electrical Installations has been written with the apprentice and trainee electrician in mind and will be of special value to students committed to NVQ assignment or project work, genuinely reflecting the real world of electrical installation engineering beyond the college lecture room door.

Graphical symbols used in illustrations are generally drawn from British Standard 3939, but where clarity is necessary, pictograms are provided and used as an alternative to provide clearer understanding. This eliminates the need to make reference to British Standards directive 3939.

Previous knowledge of the basic concepts of electrical principles would be helpful although not essential, providing the student has a practical awareness of electrical installation work.

It is generally agreed that 'hands-on' experience is a beneficial approach to learning. Regrettably mistakes are often made, but experience gained from such errors enable us to build upon our existing knowledge and skills, thus leading to a well-balanced understanding of electrical installation engineering. The knowledge gained from this book will help equip the reader with valuable insight and provide an avenue to academic advancement within his or her chosen career.

Christopher Shelton
January, 1996

Acknowledgements

I would like to thank the following for their assistance and time given in the preparation of this book: Andrea Shelton for her photographic work; Christopher Barr for illustrations 1.5, 1.7, 2.4, 2.7, 6.52 and 8.7; Steven Wooding for illustrating Fig. 6.23; Yvonne Palmer for editorial typing and Hubert Angst of Cerberus (Männedorf, Switzerland) for advice given concerning fire detection and alarm systems.

I would also like to thank the following companies and establishments for their help in the making of this book:

Britmac Electrical Limited, Broadlands of Romsey, City and Guilds of London Institute, Longcastle Limited of Preston, Osram Limited, Salisbury College, Tate Fire and Security Protection Company Limited, The Institution of Electrical Engineers.

The author and publishers are grateful to the following for permission to reproduce copyright material:

Marshall-Tufflex Limited for our Figs 7.2, 7.5–7, 7.10–12, 7.21, 7.51 & 7.65; Photain Controls Limited for our Figs 3.11, 3.18, 3.20–22, 3.25 & 3.28; Robin Electronics Limited for our Fig. 7.15; Systems & Electrical Supplies Limited for our Figs 1.8–10, 1.12, 3.42 & 5.21; Walsall Conduits Limited for our Figs 7.49 & 7.52.

Extracts from BS 3939 (our Figs 1.2, 9.1, 9.4 & 9.10 and our Table 6.10) are reproduced with permission of BSI. Complete copies can be obtained by post from BSI Customer Services, 389 Chiswick High Road, London W4 4AL; telephone 0181 996 7000.

Special thanks to my wife, Shirley, who acted as my literary adviser and typist and for her valuable support, assistance and encouragement.

Finally, I would like to thank the staff at Addison Wesley Longman Higher Education for making this book possible.

1 Industrial studies and safe working practices

In this chapter: The structure of an electrical contracting firm, the grading system, settlement of disputes. Designing an electrical installation. On-site security, working relationships and customer relations. Safe practices. Electricity at Work Regulations (1989). Potential career prospects. Choosing the right firm.

The structure of an electrical contracting firm

The internal structure of an electrical contracting firm will obviously vary from company to company depending on the number of employees and size of the business concerned. Some establishments are forced to reduce their overheads by excluding certain departments and, in doing so, find that they are dependent on others to carry out additional duties in order to protect the interests of the sections which have been cut back.

Large, and well-established public limited companies are able to organise themselves far more appropriately and provide good departmental structure.

Typically, a firm such as this would be affiliated to a trade organisation such as the *Joint Industry Board for the Electrical Contracting Industry* (JIB). A company of this calibre can offer excellent career prospects. This subject will be studied in greater detail later in this chapter when attention will be given to choosing a suitable firm.

Structure 1.1 illustrates the departmental arrangement of an average electrical contracting company while Table 1.1 profiles the roles and responsibilities of both employer and employee together with their working relationship with their client.

The Joint Industry Board for the Electrical Contracting Industry

The aims and objectives of the Joint Industry Board are principally to manage and administer the relationship between employers and employees working in the electrical contracting industry throughout most of the United Kingdom.

The Board's scope of interest also includes the following:

1. Annual holiday entitlement
2. Benefits: sickness with pay, accidental death, permanent or total disability.
3. Disputes procedure.
4. Financial incentive schemes.
5. Grading of operatives according to qualifications and ability. (This will be considered later in the chapter.)
6. Industrial determinations.
7. JIB employment pool.
8. Minimum working hours.
9. National wages, conditions and overtime.
10. National working rules.
11. Redundancy procedures.
12. Safety, health and welfare in the workplace.
13. The control and regulation of employment.
14. Travelling time and other allowances.
15. Vocational training schemes and apprenticeships.

The Joint Industry Board acts as a coordinating agency for the electrical contracting industry and is active throughout England, Wales, Ulster, the Channel Islands and the Isle of Man.

The grading system

Operatives employed by a participating JIB member company are graded according to qualifications and

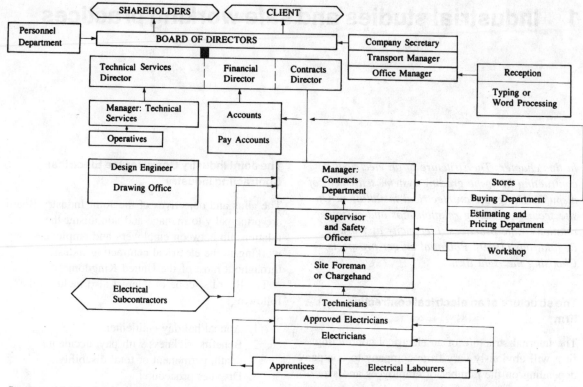

Structure 1.1. The departmental structure of a middle-sized electrical contracting company. (Can vary, according to company policy.)

experience. In summary, the four basic grades are as follows:

1. Technician
2. Approved electrician
3. Electrician
4. Labourer

Grading definitions

Technician

The following listed qualifications are required before an operative may apply for a *technician*'s grade:

1. A registered apprenticeship with practical training.
2. City and Guilds of London Institute 'C' Certificate or an NVQ, Levels 1, 2 and 3 in electrical installation engineering.
3. At least 27 years of age.
4. A minimum of five years' experience as a foreman, chargehand or approved electrician.

5. At least three years' experience in a supervisory capacity.
6. Knowledge of designing and laying out an electrical installation economically.
7. Knowledge of current working rules, regulations, British Standards and Codes of Practice.

Alternatively, an operative can be graded as a technician if he or she shows exceptional technical skill and ability beyond that which could be expected from an approved electrician.

Approved electrician

To be graded as an *approved electrician* an operative must hold the following qualifications and performance criteria:

1. A registered apprenticeship with practical training.
2. City and Guilds '236' Course, Part 2 or a National Vocational Qualification in

TABLE 1.1   The roles and responsibilities of both employer and employee

Department or rank of employee	Role	Responsibilities/duties	Status (approx.)
Board of Directors	Managing the day-to-day affairs of the company. Advisers, decision and policy makers	To remain solvent, create wealth, to protect the interests of the shareholders and client. To maintain good working relationships	1
Company secretary	Legal and financial	Legal duties. May sign documents on behalf of the company. Maintains company financial interests	1.1
Personnel department	Dealing with staff appointments and the welfare of all employees	Serving the welfare and addressing the problems of all company employees. Arranging interviews and appointments. Spokesperson for company announcements. Responsible to the Board of Directors	2
Vehicle transport manager	Supplying and maintaining company transport. Dealing with insurance matters and legal transport requirements	Ensuring that all vehicles are road worthy, are taxed, insured and Ministry of Transport tested. Responsible to the Board of Directors	2
Office manager	Department head. Provides all clerical and administrative requirements	Administrative records. Letter writing. Providing requested information from computer or files. Responsible to the Board of Directors	2
Reception	Receiving clients and visitors. Queries. Introductions. Telephone duties	Welcoming clients and visitors and to see to their immediate requirements. Often telephone and typing duties are called for. Responsible to the office manager	2.1
Word-processing department	Word processing or copy typing company documents and letters	Copy typing or word-processing duties which include letters, documents, accounts and promotional information. Responsible to the office manager	—
Accounts	Financial matters	To inspect and maintain company accounts. Keeping records. Invoicing customers. Settling suppliers accounts. Responsible to the financial director	2
Pay accounts	Wages, salaries and expenses	Ensuring that wages, salaries and expenses are paid and that other personal financial requirements such as sick pay are attended to. Responsible to the financial director	2
Contracts manager or engineer	To coordinate the needs of the client and act as a link between senior management and the workforce; to advise	Managing the contracts department and monitoring progress. To maintain agreed budget lines. To advise and recommend. Control of subordinates. Attending meetings with clients or potential clients. Maintaining the policy of the company. Responsible to the contracts director	2
Supervisor	Supervising site progress. Liaison officer between management and the workforce	Regular site visits to oversee progress of work. To discipline. To advise and inform and cater for site material requirements. General administrative duties. Responsible to the contracts manager	2.1
Site foreman	Maintaining progress and harmony at site level. Company site representative	Responsible, at site level, for all electrical work undertaken. Coordinator and technical adviser. Will liaise at site level with the architect and other companies' representatives. Responsible to the supervisor	2.2
Chargehand	Responsible for a small number of operatives of mixed grades	In charge of a small team of operatives. Non-policy decisions can be made and advice given. Will maintain production and liaise with the site foreman. Often a chargehand is responsible for his/her team's work sheets. Responsible to the site foreman	2.3
Technician	Technical overseer	Similar responsibilities as carried out by a chargehand. Can make technical decisions at a local level. Responsible to the site foreman	2.3

TABLE 1.1 continued

Department or rank of employee	Role	Responsibilities/duties	Status (approx.)
Approved electrician	A mature and fully experienced operative who is fully qualified	Carrying out tasks of a technical nature and must be thoroughly conversant with electrical installation engineering practices and problems. Responsible to the chargehand or technician	2.4
Electrician	Supplemental to an approved electrician. Often recently qualified in electrical practices	To assist and carry out tasks of which he/she is technically competent. Responsible to the chargehand or technician	2.5
Apprentice	To study and learn through *hands on* experience and by watching others carry out electrical installation work. Attainment of NVQ, Levels 1, 2 and 3	To develop and build on existing skills in order to qualify as an electrician. To carry out supervised work according to his/her technical ability and experience	2.6
Electrical labourer	Supervised non-technical electrical work. Assisting where required	General duties usually of a non-technical nature. Example: chasing walls, threading and cutting steel conduit. Drilling holes, etc.	2.7
Electrical subcontractor	Electrical duties according to grade	To assist and carry out electrical duties of which he/she is technically competent. Responsible to the contracts manager/engineer but usually, in practice, to the site foreman	—
Design engineer/ drawing office	To produce a working drawing from an idea or scheme submitted by a client	Meeting with client. Site visits. Production of draft proposals in keeping with the *Wiring Regulations*. Advising, and answering the site foreman's queries in relation to the working drawings. At times the design engineer may be asked to attend a site meeting if considered necessary. Responsible to the contracts manager	2.1
Stores	Issuing of material items such as cable, plant and accessories on production of a *Stores Requisition Order*	Supplying material items for the job. Stock taking. Administrative duties and ordering. Responsible to the contracts manager	2.1
Estimating/pricing department	Pricing enquiries relating to potential assignments or undertakings	Pricing from site visits and customer's provisional drawings or sketches. A possible formal meeting with the customer. Administration work. Responsible to the contracts manager	2.1
Buying department	To seek to obtain the best possible price for material items needed for a contract. To order material items	To negotiate preferential discount prices with wholesalers. To keep within the budget of the proposed installation. Responsible to the contracts manager and the accounts department	2.1
Workshop	Site built assemblies, etc. Access to fixed power tools. A means to carry out repairs of a delicate nature	All power tools and fixed machine tools must be maintained and inspected regularly. The workshop should be kept clean and tidy at all times and *exit* and formal *escape routes* kept free from obstructions. Often responsibility for the workshop is given to an electrical supervisor	2.2
Technical services department	Electrical maintenance, small works. Electrical breakdowns	Usually a 24-hour seven-day week service. Operative must be competent in electrical fault finding. Usually given to approved electricians or technicians.	2
Client	Customer	To provide data and relative information to produce a fair and reasonable price accompanied by accurate working drawings	—

electrical installation engineering. Levels 1, 2 and 3.

3. At least 22 years of age and a minimum of two years' experience working as an electrician.
4. Must be able to work without supervision.
5. A good working knowledge of current rules and regulations, British Standards and Codes of Practice serving the electrical installation engineering industry.

Electrician

For operatives wishing to be graded as an *electrician* the following criteria must be met:

1. Must be adequately trained and served a registered apprenticeship.
2. City and Guilds '236', Part 2 certificate or an approved equivalent National Vocational Qualification.
3. Experience and ability relative to age.
4. Must have the ability to carry out electrical installation work efficiently.
5. A working knowledge of the current *Wiring Regulations*.

Labourer

No technical or academic qualifications are required by an operative wishing to be graded as a *labourer* as all work undertaken is unskilled and supervised.

Settlement of disputes: conciliation machinery

Disputes if not resolved promptly can become prolonged and rapidly lead to stoppages in production. It is therefore essential that conciliatory action is taken immediately to avoid any possible escalation. Structure 1.2 summarises, as a flow chart, a disputes procedure adopted by JIB member companies. The Joint Industry Board recommend that all possible avenues should be sought to resolve disputes as early as possible. Prompt action is necessary at each procedural stage.

Designing an electrical installation

Table 1.2 is a simplified block diagram showing the progressive measures taken when designing an electrical installation. Some stages may be omitted altogether while others are supplemented, depending on the size and type of installation undertaken.

Let the proposed contract between client and the electrical installations company be as follows:

1. To convert a small village hall measuring 30 metres by 14 metres into a motor vehicle repair workshop.
2. Original electrical installation to be removed by others leaving a 150 amp triple-phase and neutral main switch fuse, to be used again for the new installation.
3. Earthing arrangements: 'TN–S' (by means of the supply authority's cable sheath).
4. Building work, telecommunications, fire and intruder alarm system to be carried out by others.
5. Client will act as architect and clerk of works.
6. All proposals to be tendered by the client.

Material requisites: from drawing to site
Figure 1.1 (p. 9) illustrates, using graphical symbols, the proposed electrical installation arrangements that will contribute to converting a small hall into a motor vehicle repair workshop.

In order to transfer both data and product information from a specification and working drawing into tangible requirements, a provisional material and requisites list must first be prepared. This alone will usually provide sufficient needs in order to start the installation and any outstanding items can be taken off the drawing or called for when required. Items prescribed from the drawing must be checked against the specification. This will guard against inadvertently missing out specialised considerations not included on the working drawing or preferential products asked for by the client.

Preparing a list

Using a scale rule, measure the quantity of trunking, conduit and cable that will be required for the job. Remember that an electrical location drawing is only two-dimensional and that depth will need to be taken into consideration. Figure 1.2 (p. 9) illustrates a small selection of graphical symbols drawn from the International Electrotechnical Commission and British Standards 3939.

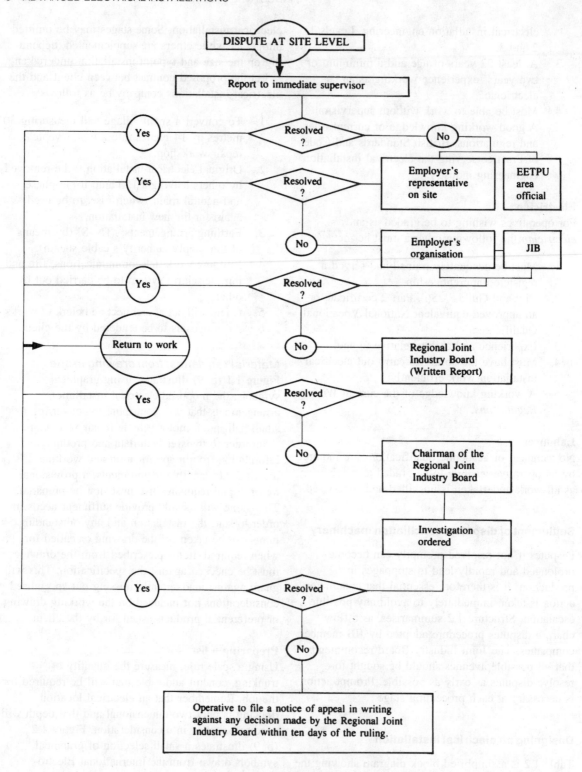

Structure 1.2. Disputes procedure adopted by JIB member companies.

TABLE 1.2   Progressive measures taken when designing an electrical installation

Step	Stage	Input by ...
1	Customer's initial proposal	Customer
2	*In-house review and survey*: Visual inspection, tests, measurements, fabric viability. Gas, water, electricity and input services identified. Size of the electrical mains checked that it is appropriate for the proposed installation. Constraints and opportunities identified. Doubts recorded for appropriate advice. How the existing installation will be affected by the new proposal. Information relating to the age and maintenance history obtained. Switchgear components checked for their effectiveness against the proposed working life of the installations. (*Example*: Motors, transformers, cable and conduit, etc.) Check existing installation that it meets requirements of the current *Wiring Regulations*. Discuss procedure for dealing with possible emergencies such as fire, acts of terrorism, toxic atmosphere and explosion, etc. Identify contract specifications.	Customer Contracts manager Head of design
3	*Formal meeting with client*: Requirements discussed in full. Review and survey discussed. Planning and legal requirements identified. Preferences on the proposed electrical installation discussed keeping in mind both planning and regulatory requirements such as building control, fire regulations, health and safety, etc. Identify regulatory and planning requirements that appear unclear such as emergency lighting and illuminated exit sign arrangements, etc. Constraints and opportunities identified. *Example*: legal, technical, financial and resources. Timetable provisionally discussed. Potential difficulties affecting the final cost of the project identified and recorded. Other factors affecting the electrical installation such as corrosion, heat, cold, moisture and loadbearing strengths discussed. Identify relevant specialist services such as nurse call, fibre optic installation, and telephones, etc. Check that these services are not outside the professional expertise of the company. Identify any legal requirements which will have influence upon the project. Check areas of the specification which conflict with the *Wiring Regulations*, local bye-laws or British Standards and refer to the authority applicable.	Customer Head of design Company estimator
4	*Assurances*: Relevant assurances to cover any foreseeable risks must be checked and confirmed. (*Example*: Employer's liability, etc.)	Contracts manager/ legal department
5	*Provisional drawings*: Simple draft working drawing produced to aid pricing the proposed contract.	Junior draftsman
6	*Cost data*: The proposed contract should be priced taking into account the following: Labour and material costs, expenses, the hiring of plant, cost-saving opportunities, factors likely to have cost implications on the proposed installation. Financial factors affecting the day-to-day running of the contract such as a site telephone, rubbish skip, portable lavatory, etc. Price calculated using provisional drawings.	Estimator Buying department Specialised companies
7	*Calculated price*: The price is offered to the customer and requirements unable to be met are identified and fully explained to the customer.	Estimator through the contracts manager
8	*Price and daywork rates approved*: Meeting with customer.	
9	*Drawings*: Working drawings produced and additional information supplied to the customer.	Drawing office
10	*Proof reading*: Proof reading by customer, including additional information relevant to the contract, proposed timetable, etc.	Customer
11	*Working drawings*: Working drawings corrected if found to be necessary.	Drawing office
12	*Proof reading*: Final check of corrected drawings made by the customer. Drawings accepted. Contract drawn up.	Client and the contracts manager
13	*Meeting with customer*: A working programme is prepared for the customer in the form of a bar chart (Figure 1.3). Potential disruptions of normal customer activity discussed. Details of co-contractors recorded. Contract signed and recorded.	Contracts manager
14	*Working drawings*: Site drawings, charts, plans and schedules printed. Job specification and bar chart programme photocopied. Plans and schedules verified by the customer.	Drawing office Customer

TABLE 1.2 continued

Step	Stage	Input by ...
15	Legal responsibilities are made clear to the job holder (the person responsible for the proposed installation). Commence work. Safety regulations implemented. Check that the site layout is safe for work to proceed. Identify any hazardous or personal risk areas. Appropriate equipment brought to site and arrangements made for its safe keeping. Final check made on site input services for compatibility with the proposed project. Site security (access gate, fences, warning notices, etc.) brought into play. Free site from obstructions to enable safe delivery of material items. Erect site accommodation (rest room, storage compound, toilet, etc.). Site office equipment provided together with telephone, drawings, plans, schedules, charts, etc. Any disruption to normal activity of the customer promptly informed. Meeting with co-contractors at site level to discuss the published timetable. Information passed to the customer to promote goodwill (disruption of power, trench digging, etc.). Accidents must be reported promptly and appropriate action taken. Customer's own *in-house* rules given to the workforce. Live circuits identified and isolated before being worked on. Progress should be monitored and opportunities to improve progress reported to an appropriate person. Monitor closely any possible causes to delay or halt progress such as strikes, material delays, sickness, slackness, etc. Record any extras, variations to the original specification. Keep to the sequence of work as programmed (see Figure 1.3). Look for alternative ways to carry out the programme of work which can offer cost and/or time benefits to your company. Where customer/co-contractor relationships are strained or complicated, an impartial negotiator should be sought.	Electrical supervisor The job holder
16	*Regular site meetings*: Information freely exchanged	Customer Contracts manager
17	*Cost data*: Cost data must be collected regularly and accurately recorded. Budget control systems should be brought into play if considered necessary.	Contracts manager/ electrical supervisor
18	*Contract completed, tested and commissioned*: Faults are reported and dealt with promptly. Snags listed and made good. Information data prepared and given to the customer. Unused material, drawings and specifications removed from site, Temporary accommodation removed from site. Disturbed fabric and structures made good. Any damage incurred during the contract to be notified to the customer. Site tidied to meet with the requirements of the agreed contract. (If this is not carried out the customer could withhold payment or take legal action.) Check that all tested connections are labelled to meet any specified demands.	Site supervisor
19	*Job formally handed over*: Test certificates given to the customer. As-fitted drawings produced. Principles and methods of safety applicable to the contract must be given to the customer. Up-to-date information on relevant regulatory and statutory requirements should be sent to the customer, preferably in written form. (*Example*: Revised British Standards, new fire regulations, etc.) This information can be obtained from journals, professional organisations and local authorities. Information and advice must be provided in a way that will generate goodwill and favour towards your company. Evaluate customer needs (additional work, technical services, improvements, emergency breakdown arrangements, etc.). Identify opportunities for improvements in which the customer will be the beneficiary.	Contracts manager
20	Prepare for the next job.	

Since not all electrical installation plans are accompanied by a relevant legend, it is wise to become familiar with the type of graphical symbols used in such drawings.

Next prepare a list in random order of all lighting, power and accessories and appliances that will be needed. This may be set out in many different ways but is best presented as a formal list written in a stores requisition duplicating book. The carbon copy will be retained as an all-time reference while the original order can be given to the in-house storeman or wholesaler.

As an example, an inventory could be drawn up in tabulated form, as shown in Table 1.3. Once signed, hand it to the person concerned and await delivery of the requested items. A list drawn up in this way is both practical and helpful; it is easy to read and reduces the possibility of error and misinterpretation occurring.

A possible alternative way   Instead of listing each material item required from the working drawing,

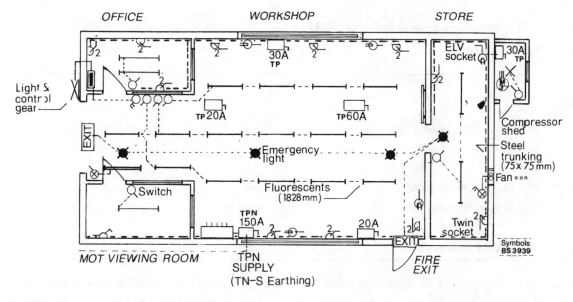

Figure 1.1. Proposed electrical installation; a motor vehicle repair workshop.

invite the company estimator to supply a photo-copy or computer printout of all the materials priced for the installation. This, regrettably, may not be welcomed as an estimator can often be extremely busy, sometimes having additional company responsibilities and will not always cooperate with such a request; but it is worth trying.

On receipt of goods

All material items arriving on site from a wholesaler or an in-house electrical store must be

Fluorescent	Fan
Emergency light	Horn
Switch	Siren
Switch–double pole	Bell
Switch with pilot light	Fuse
Wall light	Inverter
Multiple socket	LED

Figure 1.2. A small selection of graphical symbols drawn from BS 3939 and IEC.

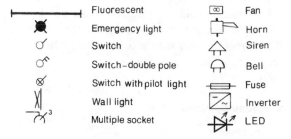

Figure 1.3. Bar chart showing the sequence of work.

TABLE 1.3   An example of a material requirements list
(Once an item has been taken off the drawing the BS 3939 symbol should be ticked with a pencil)

XEIMEA LIMITED		Material Requirements Form No. 012	
Job No. 161038		Date: 16 Oct.	
Item	Number off	Origin of manufacture	Product no.
Trunking (75 × 75 mm)	30 lengths	Walsal	T3003C7G
Trunking tees (75 × 50 mm)	3	Walsal	T3252C71G
Trunking tees (75 × 75 mm)	1	Walsal	T3053C71G
⋮			
Dust/splashproof fluorescent fittings	18	Duralite	DUR 16 SS
Twin 1828 mm fluorescent fittings	1	Speedpac	SP 26
⋮			
RCD protected socket outlets	6	Any	—

signed for and the delivery recorded. File safely away all advice notes and in-house material receipts and check thoroughly that the delivery note corresponds with items supplied. Now and again wholesalers make mistakes and it is far easier and less of an inconvenience to rectify an error or oversight at the time of delivery than at a later date. Always record deficiencies promptly and accurately.

In-house documentation

Time sheets

It is important to your company that *time sheets* are accurately completed on a day-to-day basis listing the names and addresses of clients, nature of work carried out, materials used and time spent (including travelling time) on the job. Without this information total chaos could follow.

Job sheets and daywork sheets

Job sheets are issued by the company to provide accurate and precise information concerning the job that is to be undertaken. Any additional work which is asked for and not included within the contents of the job sheet will be recorded on a *daywork sheet*. The additional work undertaken will then be charged to the customer at a standard daywork rate.

Reports

From time to time your company may ask you to carry out a formal *test of safety* on a property. This will be carried out both visually and instrumentally and results logged for future review. It might be advisable to record using a pencil and paper when making observations or carrying out formal tests. This will make it easier to present a formal report which is both articulate and legible. Filling out forms is not everyone's speciality but it plays an important role in electrical installation engineering.

On-site security and preparation

Site security is often considered to be the responsibility of others. Often it is, on very large industrial undertakings, but we all have a personal responsibility as well as being responsible to the company with whom we are employed.

Before setting up site, check that conditions are safe for work to proceed and free from physical obstructions to enable storage facilities of material items. Confirm with your company that they are fully insured against the following contingencies:

1. Wilful damage to accessories, apparatus and plant.
2. Accidental damage to plant and accessories.
3. Fire.
4. Theft of equipment and tools from a locked store.
5. Cable and accessories stolen once installed.

It is always wise to check with your company

secretary, as often there are hidden insurance contingencies to consider which might prohibit leaving company property on site overnight, regardless of whether it is securely locked away or not. Should this be the case, operatives have little choice but to return company property and personal tools to a safer keeping at the end of a working day.

When both plant and personal tools can be left on site they must be locked away in a secure containment. If a basic wooden shed is provided, check that it has an accompanying floor and that there are no windows present. This will help to hinder and frustrate a potential thief who will be unable to see inside or tunnel underneath to gain entry.

Keys
It is wise to keep two keys, one in the safe keeping of an electrician charged with supervising the project; the other placed in the care of a responsible operative.

Close and lock all doors when leaving. Make sure that the electricity has been turned off and that any oil heaters have been extinguished. Never leave a key in a convenient place such as under a brick, on a ledge, under a mat or inside a letter box. Hidden keys can be found by a would-be intruder and any insurance claim is often weakened if material items or personal effects are stolen.

Technological security
Gone are the days when a building contractor would employ a watchman to patrol the site during the hours of darkness. Today we have to approach the problem of security by use of technology. Discreetly located passive or active infrared detectors controlling peripheral security lighting and a simple intruder alarm system (see Chapter 5), accompanied with a loud industrial sounder, would help to discourage all but the determined professional thief from gaining entry to secure areas.

When a building site has no technological means of protection it can be an open invitation for an opportunist to look over the site during the hours of darkness and commit burglary. Where appropriate, site access prevention measures should be taken using gates, barriers and warning notices.

Tools
Never lend tools or equipment to site operatives; it may be the last time you will see them! Tools are expensive and it is wise to personally insure them, as usually an employer's insurance policy does not cover tools owned by his or her employees. For a reasonable yearly premium most tool kits can be insured against a number of contingencies which will help to provide peace of mind. Alternatively, the Joint Industry Board (JIB) operate an insurance scheme to compensate the value of an operative's tool kit from loss or theft while working for a JIB member company. Regrettably this is only open to graded JIB electrical workers. Operatives who are not affiliated to the Joint Industry Board would have to seek private insurance.

Many electricians place a layer of coloured insulation tape around the handle or shaft of their pocket tools (Figure 1.4). This alone will not discourage a potential thief, but it will provide instant recognition should the tool be thoughtlessly or unwittingly misplaced.

Unlocked vehicles
The loss of tools, plant, cable and accessories from unlocked vehicles is not usually covered by personal insurance. Claims will only be considered if a vehicle is locked and entry is gained by physically breaking in.

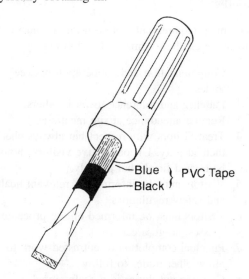

Figure 1.4. A pocket screwdriver can be quickly recognised when coloured insulation tape has been wrapped around its shaft.

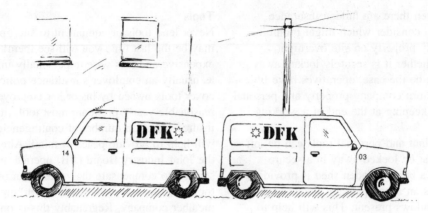

Figure 1.5. Vehicles parked back to back will help provide additional security.

It is always wise to lock vehicles parked in offical building site compounds. When two company vans are used, park them back to back (Figure 1.5), as this will provide additional security when vehicles are left overnight.

On-site working relationships

It is important to maintain a good relationship with your working companions. A smooth running site is often a happy one and this should always be an objective. Represent your company favourably,

Remember, cooperation is better than conflict.

The main requirements for securing a congenial atmosphere can be summarised as follows:

1. Communication and cooperation between trades.
2. Patience and tolerance towards others.
3. Regular attendance at site meetings.
4. Treat visitors with respect but always check their identity. Uncooperative visitors should be treated firmly but politely.
5. Visitors must be informed of relevant health and safety regulations.
6. Visitors must be informed of site procedures (passes, no-go areas, etc.).
7. Punctual completion of entrusted work to allow other trades to follow.
8. Carrying out duties in a professional manner.
9. Keeping to an agreed programme of work.
10. Regular managerial site visits.
11. Regular productivity.
12. The avoidance of long periods of absence from site.
13. Reasonable requests from colleagues should be met both promptly and willingly.
14. Avoid relying on others to provide tools or plant.
15. When work or personal difficulties occur, assistance should be provided.
16. Carrying out verbal or written site instructions issued by the general foreman or architect.
17. Taking due regard and responsibility for the property of others.
18. Personal consideration in respect to others and ethnic minorities.
19. Avoidance of noise pollution. Reduce the level of sound from radios. One person's music is another's torment!
20. Treat working colleagues in a manner which promotes goodwill.
21. If in doubt, do not be too proud to ask.
22. An input of managerial skills is required from all trades involved in the project.
23. Get to know your customer well. Remember names.
24. The punctual delivery of material items.
25. The supply of protective clothing and suitable covering for material items.
26. Advise of disruption to your customer's activity.

27. Maintain a site diary. Enter site visits, names, additional work, variations and instructions given by others.
28. A breakdown in discipline should be promptly reported.
29. Provision of suitable toilet facilities, rest room and a 110 volt electricity supply for general-purpose tools.
30. Keep a current edition of the *Wiring Regulations*.
31. A suitably qualified operative must be in charge of the installation.
32. Always respond to requests for information from visitors or other trades people.
33. Never play silly jokes on colleagues such as hiding lunch boxes, clothing, car keys, etc. This could cause bad feeling.

Forming an effective working relationship

We should try to adopt a simple personal code of ethics when dealing with customers or potential clients. A good image coupled with a high standard of professionalism is essential. It is no longer sufficient just to give the required expertise and good workmanship demanded by customers. You will not be judged on this alone. A combination of old-fashioned protocol blended with a few social graces will often be sufficient to generate the necessary confidence to win a customer over and form a good working relationship. Always explain potential causes for disruption to your customer's activity such as total or partial power disruption or when there is need to work in occupied areas.

Listed in random order are points to consider or discuss when dealing with a customer.

1. *Personal appearance*: A neglected and unkempt appearance often creates the wrong impression. Dress for the job!
2. *Courteousness*: Practise good manners and keep away from argumentative topics. Be polite and respectful to customers. Promote a good and positive relationship.
3. *Personal hygiene*: Body odours and bad breath can be offensive. Keep any unpleasant habits under strict control.
4. *Attitude*: Maintain a good business relationship with your client. Be pleasant and helpful. Show enthusiasm and interest when carrying out the task in hand but bear in mind that familiarity can breed contempt.
5. *Image*: Present an acceptable image by keeping up a good standard of workmanship. Acquaint your customer with your intended programme of work. Be professional and know your job!
6. *Tact*: Avoid gossiping about mutual business relationships. Never smoke in a customer's home; he or she might find it offensive. Trust your own judgement. Explain your programme of work including any period of disruption which might occur.
7. *Respect for property*: Care and consideration must be of primary concern when work is undertaken in a client's home or business premises. Work carried out in a client's home should not be approached in the same way as tasks performed on a building site. Consideration must be given to using dust sheets and screens and care taken to avoid damage to the customer's home. It is wise to ask your client to remove or make safe any item of value in areas where work is to be carried out. A risk-free area is beneficial for everyone as clients are usually quick to claim compensation for damage caused by carelessness. Leave the completed job both clean and tidy.
8. *Safety*: Be wary of harmful substances. Respect your customer's rules relating to safety, conduct and access.
9. *Information*: Respond to requests for information promptly. Develop a professional relationship with your customer's representative and co-contractors.
10. *Positive relationships*: Treat both your customer and co-contractors in a manner which will result in greater understanding and encourage a positive relationship.
11. *Strained relationships*: Where customer/co-contractor relationships are strained or unworkable, an impartial mediator should be sought.
12. *Unavailable information*: Prompt action must be taken when a client or co-contractor requests information that is unavailable.

13. *Communication*: Avoid speaking to the technically unaware by use of jargon (the language of a trade or profession). Your listeners will know that their language is being spoken but they will find it very difficult to understand!

14. *Site records*: It is important to keep accurate records. Log additional requests and variations which are given. Remember the names of representatives of other co-contractors. It will help to forge a positive relationship with others.

15. Keep confidential information to yourself. Avoid gossip.

16. Advise of health and safety requirements such as protective clothing, hard-hat, restricted areas, etc.

17. Identify your customer's needs accurately. Provide appropriate information and explain such advice to the non-technically aware in terms they will understand.

18. Any recommendations for improvements to an existing installation should be discussed with your customer.

19. Discuss the range of services available to your customer such as small works, maintenance, emergency call-out, etc.

20. Establish the technical awareness of your customer or co-contractor and address that person accordingly.

Safety and welfare in the workplace

Safety and welfare at work are industrial areas which many take for granted these days. However, responsibility should fall on each and every one of us to try to familiarise ourselves with current safety regulations in respect of all types of work undertaken.

Accidents can be prevented by practising vigilance and being aware of the possible hazards that exist. Think ahead and calculate the magnitude of any potential danger to yourself and others. Accidents don't just happen; they are caused by lack of care and awareness of hidden perils that are ever present within our building sites and workplaces.

Worktime misadventure is not only disruptive but contributes significantly to lost productivity and wasted production hours in our industry today.

In order to promote a safer working environment, general random points of consideration relative to electrical installation engineering have been listed in summary form. Where accidents are concerned, people must always take priority over damage to property.

1. Know your site first aid person. Get to know where your local casualty and accident unit is located.

2. Open-fronted tungsten halogen lights should not be used for temporary illumination purposes, especially in thatched roof spaces or barns containing flammable material.

3. Allowing a jigsaw to stop before removing from the workpiece will prevent the blade from shattering and causing possible personal injury.

4. Provide necessary information and support to new working colleagues arriving on site. Give them time to settle in before judgement is made.

5. A well-stocked first-aid box is essential whenever work is contracted. Provide prompt assistance in accidents.

6. Heavy wooden cable drums must always be manned. Never leave a drum free to rotate unattended.

7. Work accompanied on live mains equipment. Immediate assistance can be offered should an accident ocur.

8. Frayed or mechanically damaged flexes serving hand-held power tools must be replaced immediately. Remedial action must be taken where an unsafe work practice is recognised.

9. Regularly inspect plugs serving hand-held power tools for damaged, loose or disconnected conductors.

10. Personal injury can be caused by direct contact to 'live' or extraneous projecting parts when wrist watches, bracelets and rings are worn. It is safer to remove jewellery when work is carried out on 'live' equipment.

11. Face masks should be worn in roof spaces

lined with fibre glass or when working with or dismantling asbestos. Masks should also be worn in grain storage silos or if work is undertaken in poultry or piggery holdings. It is essential that a mask is worn when using a masonry disc cutter for chasing walls. Identify potential hazards; it could prevent future medical problems.

12. Masonry nails and nails which serve cable clips have been known to snap and fly causing personal injury when not hammered home squarely. Eye protection should be worn.

13. Loud noise and ultrasound can physically damage hearing. Suitable ear protection should be worn.

14. Protective clothing, a hard-hat and steel-capped shoes or boots are essential when working on a building site.

15. Mechanical hoist, designed to carry only plant and material items, should never be ridden.

16. Never smoke in thatched roof spaces or areas containing flammable material.

17. Get to know the location points of fire alarm call points, extinguishers and emergency exits when working in a large occupied building.

18. Employers have a legal duty to ensure that operatives are physically capable of lifting and carrying heavy loads. Be guided by the following simple directions:
    (a) Make sure that your route is clear
    (b) Chin in
    (c) Back straight and upright
    (d) Stand close to the load
    (e) Elbows tucked in
    (f) Knees bent
    (g) Grip load firmly using gloves to avoid sharp edges
    (h) Take account of the centre of gravity of the load
    (i) Position feet one in front of the other
    (j) Lift with your legs, not your back, taking the strain; remember the centre of gravity
    (k) Balance the load using both hands
    (l) Avoid sudden twisting movements

(m) Avoid injury to others when moving
(n) Take additional care when lifting items made from glass or items which are wrapped or greased.

It might seem as though an operative is rehearsing for a pantomime but if these directions help to avoid back injury then it is well worth practising.

19. When help is required to move heavy loads, elect one person to provide instructions. This will avoid confusion.

20. Blunt tools are dangerous. A well-ground screwdriver or sharpened handsaw is far safer and easier to use. Trim any metallic overhang around the impact area of a cold chisel or bolster to prevent torn and crushed metal from causing injury (Figure 1.6).

21. Use only a 110 volt supply for power tools. Construction site offices and storage areas served with a 230 volt mains supply must be protected against leakage currents to earth by means of a residual current device having a tripping current of 30 milliamps. Temporary installations must be checked and tested every three months.

22. To help prevent muscular strain, power tools must be handled correctly. Avoid over-stretching when drilling hard masonry.

23. Prefabricated mobile tower scaffolding must not be erected more than three times the minimum base width, which should not be less than 1.21 metres. Read the accompanying instruction.

24. Safety points to consider when working with a ladder are listed as follows:
    (a) Inspect for damage. Log inspection in day-diary.
    (b) Place the ladder on a firm base; 75° from the horizontal.

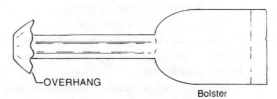

Figure 1.6. Trim any metallic overhang around the impact area of a cold chisel or bolster.

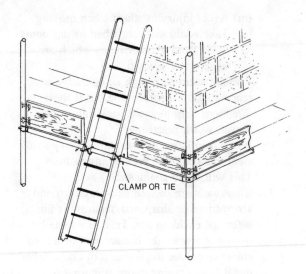

Figure 1.7. Extend the ladder 1.1 m from the top of the landing platform.

(c) Avoid suspended electricity distribution cables and telephone wires.

(d) Position the ladder correctly relative to the work to be undertaken.

(e) Do not take personal risks when climbing. Beware of grease or ice on the rungs.

(f) Extend the ladder approximately 1.1 metres from the top of the landing area, as illustrated in Figure 1.7.

(g) Lash both top and bottom of the ladder to the surrounding infrastructure if possible.

(h) One person per ladder at any one time. Face the ladder when ascending or descending

(i) Avoid over-reaching. If necessary reposition the ladder to obtain greater accessibility.

(j) Assign a colleague to stand on the foot of the ladder if lashing to the infrastructure is impractical.

25. Report potential danger areas and property defects in writing to an appropriate authority before work begins.

26. A site telephone or means of communication is essential.

27. Suitable toilet and eating arrangements are necessary for the health and welfare of operatives.

28. Always test to make sure that equipment is safe to work on. Read the safety instructions relating to equipment.

29. Unwind extension power leads fully to prevent overheating from occurring.

30. Use only enclosed-element-type heating appliances in site huts. These are best fixed to the wall supported by a suitable heat-resistant backing.

31. Fit a mains-operated, battery back-up, smoke alarm in storage areas.

32. Notify the emergency services (police, fire, ambulance), should a major incident occur. Bring into play site emergency systems such as an audible alarm for alerting or evacuating the site. Site emergencies could include: fire, toxic atmosphere, terrorist activity or explosion.

33. Suspend work immediately in the event of a fire alarm. Isolate the site power supply if it is safe to do so and proceed to the approved assembly point. Check that all staff are present. Inspect for damage to property and new installation after carrying out first aid to injured persons. Report the incident in writing.

34. An emergency can be described as an unexpected occurrence requiring immediate attention, whereas an accident may be defined as an unforeseen or unexpected event; a personal mishap.

35. It is important to be familiar with site evacuation procedures when work is carried out on large sites.

36. Site defects affecting the health and safety of operatives must be reported to the appropriate person. This could be the site manager, agent or foreman.

37. Check that the site is safe before work commences. Choose a safe means of access from the public highway.

38. Inspect that security arrangements are adequate (fencing, access gates, locks, windows, etc.).

39. Inspect hoists, mobile towers, ladders and scaffolding for defects. (*Example*: Badly

constructed hoists, damaged mobile tower sections, cracked or broken runged ladders and bent or badly constructed scaffolding.)

40. Identify site input services such as gas, water, electricity, telecommunications, steam and compressed air, etc. Check against work instructions and safety procedures. Advise if different.

Safe practices: the health and safety law

Employers have a legal duty to protect their staff and acquaint them with the demands and requirements laid down by the *Health and Safety at Work etc. Act, 1974.*

Health and safety are of prime importance to every one of us as we are not only responsible to ourselves but also to our colleagues and others.

The following summary briefly explains the principal aims and objectives of the *Act*; but it must never be considered as a complete work. Hopefully it will provide insight into the many potential hazards which may beset us when working in an industrial environment.

Legal duties and responsibilities of an employee

1. You must cooperate fully with your employer on all aspects of health and safety.
2. Refer to your safety officer or senior member of management if you have good reason to feel that a health and safety problem exists at work.
3. If a problem cannot be resolved favourably at local level and professional advice is required, the *Health and Safety Inspector* can provide suitable means for arbitration (an accepted means of judgement between two parties).

   Health problems associated with work are managed by the *Health and Safety Excutives Employment Medical Advisory Service*; the address of which may be found in the telephone book.
4. An employee must not mistreat or interfere with anything contributed by the employer which is for the health, safety and welfare of the company.
5. Operatives must be responsible for their own

health and the the health of professional colleagues who would be physically affected by an undertaking, decision or omission made by the operative.

6. Accidents must be reported promptly and recorded in the company or site accident book.

The statutory responsibilities and duties of an employer

1. When an employer employs five or more employees he or she must prepare a written health and safety policy statement and bring it to the attention of his or her operatives.
2. An employer is required by law to ensure that all workplaces are safe and without risk to health.
3. Protective clothing and equipment required by the health and safety law must be provided free of charge by the employer. For certain types of work, eye protection must also be made available.
4. An employer is obliged to form a safety committee if requested in writing by two or more safety spokespersons. He or she is also obliged to instruct and train concerning matters for the health and safety of his or her workforce.
5. Adequate fire-fighting equipment and means of escape must be provided by an employer.
6. An employer must provide satisfactory means of protection from dangerous and moving parts associated with workshop machinery. Guards, screens, fences and instructive working notices must be fitted wherever applicable.
7. It is an employer's duty to provide suitable welfare amenities. Examples may be drawn from the following:
   (a) Rest room for meals and breaks
   (b) Adequate heating
   (c) Drying room/cloakroom facilities
   (d) Fresh water for drinking.
8. An employer must provide adequate training and supervision for apprentices and unskilled personnel before allowing them to use dangerous machinery.
9. Washing, sanitation facilities and clean fresh

drinking water must be provided and maintained by an employer.

10. An employer must maintain adequate levels of illumination in workshops. Good ventilation is also required together with a reasonable ambient temperature. Workshops must not be overcrowded.

11. Adequate first aid equipment must be provided by an employer.

12. An employer must report any dangerous occurrences, certain types of injuries and diseases such as asbestosis to the enforcing agency. An accident book to formally record misadventure must also be provided.

13. The maintenance of ladders and steps are the legal responsibility of an employer.

14. An employer must not permit employees to carry or lift heavy loads that could cause the employees injury. Mechanical handling aids must be made available for moving heavy goods.

15. An employer has a legal duty to ensure that floors, stairs and passageways are erected in a professional manner and are properly maintained.

16. Adequate safety measures must be provided by an employer before allowing an operative to work within a confined space.

17. Employers are responsible for making certain that material items are moved, stored and used without risk of personal injury to the operative. All dangerous materials must be properly identified.

18. An employer must take adequate means to protect employees against the dangers of radiation and substances which could prove harmful to their health, especially in confined spaces. (*Example:* Certain types of PVC adhesive.)

19. Workplaces must be kept clean and free from hazards by an employer.

20. An employer has a duty to keep noise, fumes and dust under control.

Safety signs

From 1 January 1986 it became a legal requirement that all safety signs must comply with British

TABLE 1.4   A selection of British Standards relating to the electrical installation engineering industry. (Ideal for NVQ assignment and project work.)

Specification No.	Subject matter
BS 31	Steel conduit and fittings (imperial)
BS 88	Cartridge fuses (voltages up to 1 kV AC; 1.5 kV DC)
BS 476	Thermal combustibility factors
BS 599	High-voltage neon lighting installations
BS 1362	Cartridge fuses used in conjunction with 13 A plugs
BS 1363	Switched and unswitched socket outlets. Plugs rated at 13 amp
BS 1710	Identification of pipelines
BS 1852	Alphanumerically coded resistors used in electronics
BS 3036	Semi-enclosed (rewirable) fuses up to 100 A at 240 volts
BS 3535	Safety isolating transformers
BS 3871	Miniature and moulded case circuit breakers
BS 3939	Graphical symbols for electrical and electronic drawings
BS 4293	Residual current operated circuit breakers (RCD)
BS 4607	Non-metallic (PVCu) conduit and fittings
BS 4678	Cable trunking
BS 4752	Switches and control gear up to 1 kV AC; 1.2 kV DC
BS 4941	Starters for electric motors up to 1 kV AC; 1.2 kV DC
BS 5266	Emergency lighting
BS 5378	Graphical safety signs
BS 5467	Armoured cables with thermosetting insulation
BS 5733	General requirements for electrical accessories
BS 5839	Fire detection and alarm systems in buildings
BS 6004	PVC-insulated cables for electric power and lighting
BS 6081	Terminations for mineral-insulated cables
BS 6207	Mineral-insulated cables (MI cables)
BS 6977	Insulated flexible cables for flexible connections
BS 7671	*IEE Wiring Regulations*

Standards 5378, Part 1 (see Table 1.4). This ruling resulted from a European Community directive in 1977 and the *Safety Signs Regulations* of 1980.

Safety signs are divided into four categories of which each has its own shape and colour. They are designed to provide instant recognition as to the type of industrial territory an operative is approaching. A simple silhouetted symbol conveying an instant safety message, without the need for words, is located in the centre of each sign.

Safety signs are divided into four groups:

1. WARNING, *of which there are nineteen*
2. PROHIBITION, *of which there are eight*
3. MANDATORY, *of which there are fourteen*
4. EMERGENCY/SAFE CONDITION, *of which there are seven.*

Warning signs Designed in the form of a black-bordered triangle, an easy-to-read symbol is silhouetted on a yellow background (Figure 1.8). Signs of this type are used to warn of potentially hazardous conditions such as risk of electric shock, exposure to radiation, the risk of explosion or fire, etc.

Prohibition signs This category of sign is disc shaped and supports a red border with a 45° crossbar over a silhouetted black symbol placed on a white background, as shown in Figure 1.9. Prohibition signs advise of no smoking areas; where water is unfit to drink; places where ladders are prohibited and where electrical appliances must not be switched off. In all there are eight different types, each conveying its own unworded message.

Mandatory (compulsory) signs These are designed as a white symbol on a blue discoid-shaped background, as illustrated in Figure 1.10. This type of sign is employed where, for example, ear, eyes and head protection is demanded.

Emergency/safe condition Designed in the shape of a rectangle or square, an easy to recognise white symbol is silhouetted on a green background, as depicted in Figure 1.11. This style of safety sign is used to indicate areas of safety such as assembly points, emergency exits, smoking areas and first aid posts.

Figure 1.12 illustrates a selection of safety signs adopted by BS 5378, Parts 1 and 3.

Electricity at Work Regulations (1989)

Electricity at Work Regulations are statutory (legal) requirements affecting both employer and employee alike.

Black symbol & border   Yellow background

Figure 1.8. 'Danger — electricity' sign to BS 5378.

Black symbol

White background

Red border

Figure 1.9. 'No smoking' sign to BS 5378.

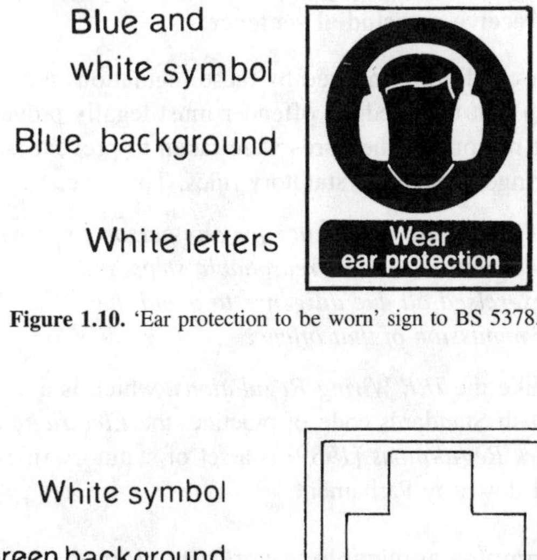

Blue and white symbol

Blue background

White letters

Wear ear protection

Figure 1.10. 'Ear protection to be worn' sign to BS 5378.

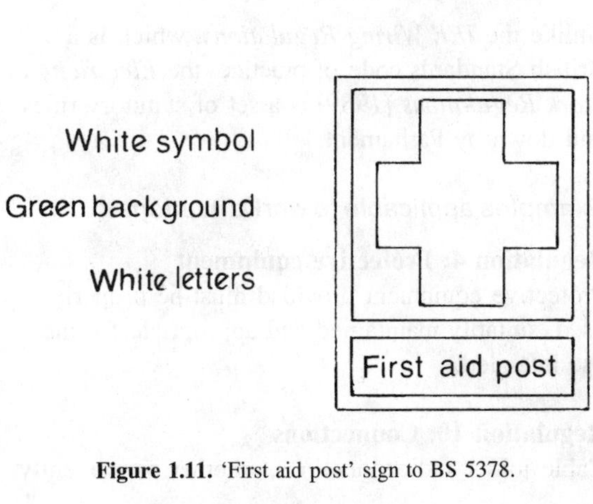

White symbol

Green background

White letters

First aid post

Figure 1.11. 'First aid post' sign to BS 5378.

PROHIBITION (Red)

WARNING (Yellow)

COMPULSORY (Blue)

SAFE CONDITION (Green)

Figure 1.12. A selection of safety signs meeting the requirements of BS 5378, Parts 1 and 3.

Should an operative or employer infringe these demands the guilty party could be heavily fined or receive a custodial sentence.

The conditions imposed by these regulations are such that the alledged offender must legally prove that reasonable measures were taken to prevent an infringement of the statutory rules. To quote:

> ... *it shall be a defence for any person to prove that he/she took all reasonable steps and exercised all due diligence to avoid the commission of that offence.*

Unlike the *IEE Wiring Regulations*, which is a British Standards code of practice, the *Electricity at Work Regulations (1989)* is a set of statutory rules laid down by Parliament.

Examples applicable to work situations

Regulation 4: Protective equipment
Protective equipment provided must be properly used, suitably maintained and appropriate for the use intended.

Regulation 10: Connections
Cable joints and connections, whether permanently

or temporarily made, must be both electrically and mechanically suited and relevant to the installation so that danger can be prevented.

Regulation 11: Means for protection against excess current
Circuits must be efficiently protected against excess current caused by short circuit or over-current and be positioned in a suitable location.

Regulation 13: Precautions for work on equipment made dead
Appropriate safety measures must be selected to avoid equipment which has been previously made 'dead' from becoming electrically charged, if by doing so it would cause a dangerous situation.

Further reading
The *Electricity at Work Regulations (1989)*, together with the *Health and Safety at Work etc. Act (1974)*, were established to help provide umbrella protection at the workplace.

We must all attempt to acquaint ourselves with these regulations and endorse the requirements of these two important legislative works from which we shall all gain considerable useful knowledge.

Useful publications concerning both health and safety at work include the following which are available from Her Majesty's Stationery Office, Duke Street, Norwich, or through any government bookshop.

- *The Essentials of Health and Safety at Work*
- *Safety Representatives and Safety Committees (1977)*
- *Notification of Accidents and General Occurrences Regulations (1980)*
- *A Guide to the Health and Safety at Work etc. Act (1974)*

The following publication is available from the Publications Department of the Trades Union Congress (TUC):

- *The TUC Guide to Health and Safety at Work*

Potential career prospects: choosing the right firm

The difference between working for a badly run or well run company can have a considerable effect on the well-being and future prospects of an employee. So the importance of finding out about employment conditions must take priority when applying to a potential employer.

Small companies

There are several factors to be borne in mind when considering employment with a small firm. For example, a contractor employing up to ten operatives may be totally dominant, applying extraordinarily flexible rules which change at his or her whim. This contractor would be unlikely to be a member of the Joint Industry Board for the Electrical Contracting Industry (JIB), and may not offer contracts of employment to employees. In such a firm, equally qualified operatives may be working at different rates of pay and might be expected to work overtime for basic rates. There also may be infrequent and irregular pay awards. To add to this list of wrong doings, protective clothing may not be provided when required. In order to complete a contract quickly, health and safety regulations might be flaunted. During times of recession badly run companies are particularly

prone to setting aside reasonable employment conditions in order to obtain good profits. These firms may offer poor training facilities and virtually no chances of promotion.

However, it must be noted that some small companies are extremely well run but great care should be taken to distinguish between good and bad. It is helpful to ascertain whether the employer is a member of the JIB or any other trade organisation.

Larger companies

Large private or plc electrical contracting companies affiliated to the Joint Industry Board for the Electrical Contracting Industry are an excellent choice as prospective employers. JIB operatives are graded as follows:

- Technician
- Approved electrician
- Electrician
- Apprentice
- Labourer

The grading system has been dealt with in more detail previously in this chapter. Each of these grades carries its own rate of pay and responsibilities.

One of the many advantages to be gained from working for a JIB affiliated company is that regular pay awards are set, making any need for an employee to negotiate pay with his or her employer unnecessary. The JIB also sets out clearly defined rules, regulations and a scale of allowances for travel and lodging, etc. All employees automatically become members of the *Electrical Electronic Telecommunication and Plumbing Union*. Also if the employer makes JIB combined benefit stamp contributions to the JIB/*British United Provident Association* (BUPA) *Health Scheme*, the employees are able to take advantage of BUPA health care when necessary. Other benefits may include life assurance, accidental death and permanent and total disability benefits.

Various courses which operatives may attend are organised by the JIB. Current information on various subjects is sent by post to all members. Finally, the JIB provides an avenue for grievances should disputes arise between employers and employees. This area has been fully covered earlier in this chapter.

Career options

There are many career avenues which may be considered once an apprentice is qualified, and the choice can be difficult. Often additional vocational qualifications are required (NVQs, City and Guilds, National or Higher National certificate, etc.), so it is wise to decide which career option to follow as early as possible so that the appropriate path may be taken.

Contracting

Working for an employer in the electrical contracting industry involves general electrical installation work. However, if an operative prefers not to be employed continuously by one firm it is possible to become a freelance electrical worker, better known as a 'subby' or an electrical subcontractor. This gives an electrician more independence as well as the necessity to negotiate both pay and conditions of each job he or she undertakes.

Maintenance

Another alternative is to become a maintenance electrician working, for example, in industry or perhaps in a hospital.

Specialised electrical engineering

Specialised skills are required in areas such as radio, television and the theatre or in the petrochemical or aircraft industry. Electrical installation work may also be contracted on ships or oil rigs in marine engineering.

Management

If the managerial side of electrical installation engineering is preferred, training may be undertaken to become a supervisor or foreman whose duties have been described previously in this chapter. Alternatively, it is possible to become part of a management team but additional vocational training is often required for such a position.

Technical education or representation

Teaching is another option to be considered. It is possible to become a tutor in a college of technology, training centre or manufacturer's school. Again additional vocational qualifications are needed for a post of this kind.

On the other hand, preference may be inclined towards becoming a technical consultant. This position involves providing technical advice and helping to develop new concepts.

Another idea is training to become a technical sales representative and demonstrator undertaking high-tec sales to companies or colleges.

If authorship appeals, it is possible to eventually become a technical author writing specialised papers and textbooks.

Proprietor (owner of a small business)

Finally, ambitions may extend in the direction of becoming the owner of an electrical company or perhaps becoming a company director. With the right sort of motivation and determination it is often possible to achieve the seemingly impossible ambition.

Summary

1. Well-established companies are able to organise themselves efficiently and provide good departmental structure. They are often affiliated to a trade organisation such as the Joint Industry Board for the Electrical Contracting Industry.
2. The aims and objectives of the Joint Industry Board are to manage and adminster the relationship between employers and employees.
3. Operatives employed by participating Joint Industry Board member companies are graded according to qualifications and experience.
4. Table 1.2 outlines progressive measures taken when designing an electrical installation.
5. Transfer both data and product information from the specification and working drawing to form a tangible list of requirements.
6. Sign for and log all material items arriving on site from an electrical wholesaler.
7. We all have a personal responsibility for on-site security.
8. Never lend tools or equipment to unfamiliar site operatives. It may be the last time you will see either of them.
9. Maintain a good relationship with others who

work alongside you.

10. Try to establish a simple professional code of ethics when dealing with customers or potential clients.

11. Become familiar with current safety regulations in respect of work to be undertaken.

12. Accidents can often be prevented by practising vigilance and being aware of possible and potential hazards that exist.

13. Employers have a statutory duty to protect their staff and to acquaint them with the requirements laid down by the the *Health and Safety at Work etc. Act (1974)*.

14. Employees have legal duties and responsibilities to their employer.

15. All safety signs are divided into four categories,
    Warning
    Prohibition
    Mandatory
    Emergency/safe condition
    and comply with the requirements laid down in BS 5378, Part 1.

16. *The Electricity at Work Regulations* is a statutory requirement which affects both employer and employee alike. Offenders, if convicted, can be heavily fined or receive a prison sentence.

17. Further reading:
    Electricity at Work Regulations (1989)
    Health and Safety at Work etc. Act (1974)
    The Essentials of Health and Safety at Work
    Safety Representatives and Safety Committees (1977)
    Notification of Accidents and General Occurrences Regulations (1980)

18. Care should be taken when choosing an employer.

19. There are advantages in choosing a JIB member company to work for: regular pay awards, health care, life assurance, other financial benefits, technical courses available, etc.

20. Companies affiliated to the Joint Industry Board are an excellent choice as a prospective employer.

21. Once qualified, there are many career avenues and options which may be considered:
    (a) General electrical installation engineering work
    (b) Freelance electrical work
    (c) Electrical maintenance work
    (d) Specialised work in radio, television and theatre
    (e) Petrochemical, mining and aircraft engineering
    (f) Supervisory work
    (g) Management
    (h) Technical education
    (i) Consultancy work
    (j) Representative or demonstrator
    (k) Technical writing
    (l) Technical director
    (m) Owner of a small business.

22. Structure 1.2 summarises the disputes procedure as adopted by JIB member companies.

23. Table 1.1 profiles the roles and responsibilities of both employer and employee together with their working relationship with their client.

24. Structure 1.1 illustrates the departmental structure of an average medium-sized electrical contracting company.

25. Never leave site accommodation keys in a convenient place such as under a house brick, on top of a high ledge, under a mat or inside a letter box attached to a string. Keep two sets of keys, each with a responsible person.

Review questions

1. Describe what is meant by the term, 'departmental structure'.

2. What do the initials JIB stand for?

3. State the principal aims of the JIB.

4. Name the four basic grades adopted by the JIB for electrical workers.

5. Describe, briefly, the role of a contracts manager.

6. Suggest two practical ways to transfer both data and product information from a specification and working drawing to the workplace.

7. Where should on-site accommodation keys be kept?

8. Describe a simple method to identify personal tools.

9. List three examples of entries in a typical on-site day diary.
10. Why is personal appearance important when dealing with a client or potential customer?
11. In general terms, how can accidents be prevented or reduced?
12. BS 5378, Part 1, safety signs are divided into four basic categories. List these categories.
13. Describe briefly, a *first aid post sign* designed to BS 5378, Part 1.
14. Provide three reasons why a JIB member company is often considered to be a wise choice of employment.
15. Briefly list the responsibilities of a JIB graded *approved electrician.*

Handy hints

1. Remove personal effects and power tools from your company vehicle each night. Be security minded.
2. Professionally prepared working drawings occasionally contain errors. If in doubt, seek clarification from your site manager or general foreman. Guidelines can also be obtained from the *Wiring Regulations, IEE On-Site Guide*, project specifications, manufacturer's information and work instructions.
3. Material defects can be caused by faulty manufacture, corrosion and distortion caused by heat (*example*: PVC conduit). Dispose of excess or waste material in an appropriate manner.
4. Refractory bricks must be kept dry. Wet or damp bricks will steam and cause damage to wall decorations when the storage heater is energised.
5. A nylon carpet has a low melting point. Take care not to allow a hot masonry drill-bit to come into direct contact with any nylon furnishings.
6. Each year there are about five fatal accidents involving portable electrical appliances. Portable appliance testing must be carried out with the same concern and consideration as any other professional task we undertake. To ensure a proper evaluation, each appliance must be placed and tested in the 'ON' mode. All test results must be formally recorded.
7. Domestic and industrial voltage values were changed from 415/240 volts, ±4 per cent, to 400/230 volts, +10 per cent, −6 per cent, on 1 January 1995. Voltage tolerance levels will be further adjusted to ±10 per cent by the year 2003.

2 The nature of electricity

In this chapter: Historical introduction. Atomic theory. Stactic electricity. The magnetic phenomenon: the nature of electricity. Simple generator. The root mean square value of an alternating voltage. Electricity from magnetism. Transformers. The effects of the introduction of electricity to the home. Power distribution. Advantages of high-voltage power transmission.

Historical introduction

In about 600 BC a Greek philosopher called *Thales of Miletus* discovered that by rubbing a stick of amber with a dry fur the amber would magically attract small segments of leaves and tiny pieces of straw. Unbeknown to him he had unwittingly discovered electricity.

Sealing wax rubbed on fur, glass on silk or a plastic comb run through clean dry hair will also produce this phenomenon. This type of electricity is known as *triboelectricity*, or sometimes refered to as frictional electricity. At its mildest a small electric shock can be experienced, for example, when walking across a clean dry nylon carpet while wearing shoes with synthetic rubber soles. Touching an electrical earth will release the stored energy thus producing a mild electric shock. At its worst, triboelectricity can kill and destroy by electrical energy generated in the clouds during a thunderstorm. This may amount to many millions of static volts and is very dangerous.

Nearly 2000 years after Thales discovery, a Colchester doctor named *William Gilbert* (1540–1603; see Appendix A) coined the term ELECTRICITY from the Greek word *elektron*, meaning amber. In 1600 Gilbert published a paper concerning magnetic bodies called *De Magnete* which paved the way for many future pioneering experiments with electricity.

Appendix A contains a brief tabulated account of progress and achievements in electrical science throughout history.

Atomic theory

To understand the nature of electricity, the atom must first be taken into consideration. An atom can be defined as the smallest fragment of an element that is capable of engaging in a chemical reaction. It can be subdivided into three basic principal subordinate parts.

1. *Proton*: a positively charged elementary particle with a mass 1836.12 times greater than an electron.
2. *Neutron*: an elementary particle that is neither positively nor negatively charged, supporting a mass a little more than its accompanying proton.
3. *Electron*: a negatively charged elementary particle with a mass just 1/1836 of its parent proton.

The simplest and smallest atom is that of *hydrogen* which, as Figure 2.1 shows, has just one proton accompanied by a single orbiting electron.

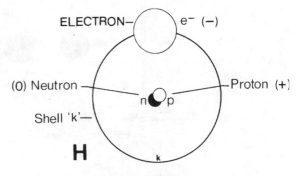

Figure 2.1. An atom of hydrogen.

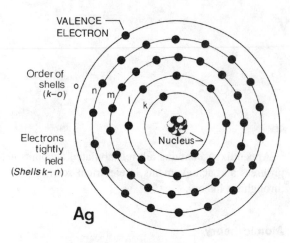

Figure 2.2. An atom of silver.

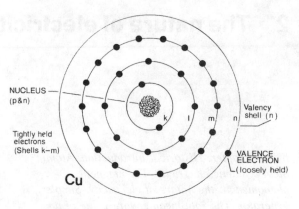

Figure 2.3. An atom of copper.

Not all elements are forged in the same proportions. Silver, for example, supports no fewer than 47 protons and 61 neutrons to form the central nucleus. This is balanced by 47 orbital electrons travelling around the nucleus at various levels. Figure 2.2 illustrates this concept in pictorial form. Copper, a lighter element than silver, has just 29 orbiting electrons and is electrically equalised by a matched quantity of protons. Clustered with the protons are 34 neutrons; together they form the nucleus (Figure 2.3).

Particle behaviour

Two basic principles governing the behaviour of subatomic particles which will be of interest to the electrical student are:

1. **Like charges repel one another; unlike charges attract one another.**

Attraction may be successfully demonstrated by a simple, but classic experiment using a tiny laboratory *pith ball* suspended by an insulated silk thread (Figure 2.4). When a hard rubber rod which has been vigorously rubbed with dry clean fur is placed in close proximity to the pith ball, the

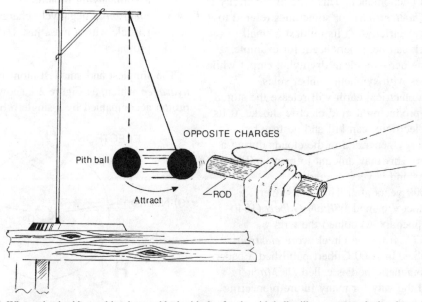

Figure 2.4. When a hard rubber rod has been rubbed with dry fur the pith ball will move towards the charged rod.

suspended ball will move towards the charged rod. Likewise, repulsion can be proved by suspending a minute hard rubber ball that has been rubbed with fur and placing a charged rod in close proximity to the ball. As both rod and ball support the same electrical charge, the suspended ball will be repelled by the hard rubber rod.

The second principle that should be taken into account is the physical law governing centrifugal force:

2. **The faster an object travels around a central body, the greater the centrifugal force becomes.**

It is this force, together with a mutual attraction between the nucleus and orbiting electrons, that prevents the atom from being destroyed by its negatively charged electrons tumbling into the positively charged nucleus!

Atomic quanta

Electrons surrounding a nucleus are arranged in *shells* or *quanta* which are internationally recognised by a single lower case (small) letter of the Roman alphabet, as illustrated in Figure 2.5. Each shell has a maximum number of electrons it may accommodate.

The shell essential to current flow is outermost and known as the *valence shell*. It may house up to eight loosely held electrons. All other electrons are firmly fixed within their respective orbits and play

no part in current flow. Both *silver* and *copper* have just one valence electron and are excellent conductors of electricity. However, the elements *silicon* and *germanium* both have four and are far less efficient conductors.

Although copper and silver both have a valence of one, silver is a much better conductor than copper. This is because an atom of silver is far larger than an atom of copper so the outer valence shell will be much further away from the nucleus and orbiting at a faster rate. Compare a bicycle in motion fitted with different diameter wheels. Although travelling at a uniform speed, the tyre provided for the larger wheel will be travelling faster than the tyre fitted to its smaller companion. Should a valence electron of silver be knocked from its orbit, the velocity at which it is travelling around the nucleus will provide for additional energy. Therefore it may be said that silver is better than copper as a conductor of electricity.

Current flow

Pressurised by an electromotive force, generated through an arrangement of conductors cutting across a strong magnetic field at right angles, loosely held valence electrons are forced from their regular orbits at high speeds. Once free, an electron will target another valence electron serving a neighbouring atom (Figure 2.6). The colliding electron provides sufficient kinetic energy for the passive electron to be knocked from orbit but then stops and joins the adopted atom. (The same phenomenon occurs when striking a stationary

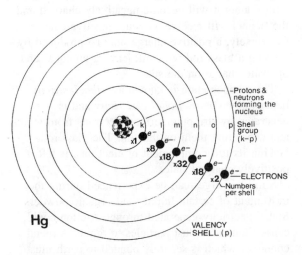

Figure 2.5. An atom of mercury.

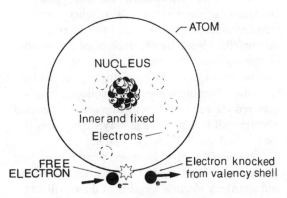

Figure 2.6. An electron targeting a valence electron serving a neighbouring atom.

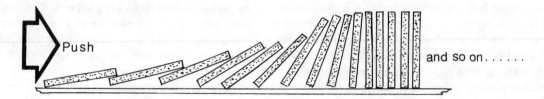

Figure 2.7. The domino effect.

snooker ball with a cue ball. The cue ball collides with the stationary ball and stops while the targeted ball is propelled forward by the energy received from the cue ball.) The buffeted particle then targets another valence electron serving an adjacent atom, and the process continues. This is current flow at its most basic level; simply a stream of negatively charged electrons pressurised by an electromotive force generated by a conductor cutting through a strong magnetic field at right angles. When considering current think of the *domino effect*, Figure 2.7.

Heat generated

When a striking electron hits a stationary electron the majority of energy is transferred from one particle to the other. However, a small amount of energy is surrendered and converted to heat. Too much current flow, especially in a small-diameter conductor, will give rise to an increase in temperature, bringing practical problems to the circuit.

Insulators

Elements such as mercury or nickel supporting a valence of 2 are, theoretically, not such good conductors of electricity as silver. When a free electron strikes a valence shell of an atom of mercury the kinetic transference of energy to other mercury valence electrons is halved. An example may be drawn from aiming a single glass marble at two others forming a row. Once collision has occurred the cue marble will stop and the targeted marbles will then proceed at very slightly less than the original speed of the cue marble.

Compounds such as glass and plastic, compared to copper or silver, contain many valence electrons and a striking electron targeted at these will add very little to current flow as energy levels will have to be shared several times. Think of our imaginary

TABLE 2.1   Insulators — a comparison of good and poor

Good insulators	Poor insulators/poor conductors
Mica	Cotton
Paraffin wax	Germanium
Plastic	Paper
Polyvinyl chloride (PVC)	Selenium
Porcelain	Silicon
Quartz	Wood
Shellac	Wool
Silk	
Sulphur	

marble hitting not two but eight target marbles!

Materials which have many electrons within their outer shell are known as *insulators* and are bad conductors of electricity. Table 2.1 provides a small selection of common insulators, comparing them with materials which are both poor conductors and poor insulators.

Static electricity

When a stick of sealing wax is vigorously rubbed with flannel it will become negatively charged and the flannel will receive an equal positive charge. Conversely, a positive charge may be produced by rubbing a glass rod with silk leaving an equal and opposite charge on the silk.

This phenomenon is known as *triboelectricity*, better known as *frictional electricity* and is generated by stripping surface electrons from atoms and depositing them on the silk. The glass rod, lacking its full complement of electrons, is left in a positively charged state.

Similar charges can be experienced during the movement of certain liquids and organic powders. Static electricity generated through friction could cause flammable dust particles to ignite. A metal container which is securely bonded to earth must always be used when pouring organic powders or flammable liquids as a plastic container could become electrically charged.

Electrical charges

Electrical charges can be compared to the physical properties of magnetic bodies in that like charges will repel one another while unlike will attract. Figure 2.8 illustrates this well-known phenomenon using bar magnets.

An electrical charge is identified by its polarity and can be recognised by the mathematical signs, plus and minus (+, −). The plus sign is reserved for bodies which are positively charged; the minus sign applies to those which are negatively charged.

A positively charged body, that is, a body without its full complement of electrons, will disturb the stability of electrons in a nearby neutral mass. If placed near to a glass insulated rod, a body possessing a positive charge will attract loosely held surface electrons from the tip of the rod. In doing so, a negative charge will develop on the opposite end. Once the positively charged source has been removed the rod will return to a normal neutral state, being neither positively nor negatively charged.

Identical charges

When two identical masses both charged with the same electrical potential but of different values are made common, the mass with the greater potential will surrender a portion of its charge to the mass supporting the lesser charge. This effect will continue until both charges are equal in polarity and magnitude. This concept may be understood more clearly by imagining two identical open-topped waterbutts made common with a small pipe fitted with a closed stop cock in line above their respective base lines. One tank is filled to the top with water while the other is only half filled, Figure 2.9. Turning the stop cock to its open position will enable the water contained in both tanks to find a common level. The pressures in each tank will then be equal.

Electrical pressure

Current flow is a physical movement of negatively charged subatomic particles. The prime mover responsible for this phenomenon is known as the

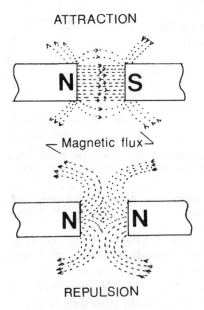

Figure 2.8. Like poles repel, while unlike poles attract.

electromotive force (e.m.f.), and is measured in *volts*. The volt is the unit of *potential difference* and is loosely defined as the difference in potential between two conductors carrying a uniform current of 1 amp while dissipating power to the value of 1 watt.

However, the *potential value of an electric charge* can be defined as the force which attracts divided and opposed charges to each other. This may be measured by means of an instrument called an *electroscope* which is able to determine electrostatic voltage differences without drawing current from the source.

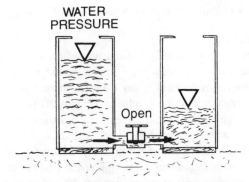

Figure 2.9. Turning the stop cock to its open position will enable the water contained in both tanks to find a common level.

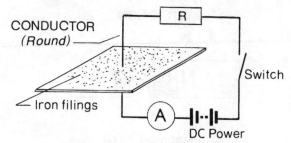

Figure 2.10. Iron filings are sprinkled over a square of 'insulating material'. (A, ammeter; R, resistor.)

The magnetic phenomenon: the nature of electricity

Oersted discovered that electricity and magnetism are related phenomena and that current flowing in a conductor can be physically shown to emit an invisible magnetic field. This may be demonstrated by routeing a vertically positioned loadbearing conductor through the centre of a small square of dry insulating material which has been lightly sprinkled with iron filing dust (Figure 2.10). The deposited dust will adopt a multi-circular pattern once the current has been switched on. The structure of the magnetic circles will be determined by the physical shape of the conductor, as illustrated in Figure 2.11. Another method of demonstrating this electromagnetic effect is by placing a simple pivoted magnetic compass needle above a loadbearing conductor in which current flow has been designed to travel from right to left, Figure 2.12. The North-seeking component of the compass needle will deflect in an anticlockwise direction once the circuit is made. Conversely, reversing the flow of current will gyrate the needle in a clockwise direction as demonstrated in Figure 2.13.

Induction

Faraday (1791–1867) realised that when a conductor is moved at right angles across a North–South magnetic field a tiny electrical potential is generated within the conductor (Figure 2.14).

A conductor, forming a simple circuit and moved across a strong magnectic field, will cause loosely held valence electrons within the conductor to shift from their regular orbits at high speed and target other valence electrons serving neighbouring atoms, *ad infinitum.*

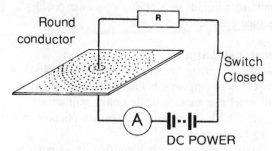

Figure 2.11. The deposited filings will adopt a multi-circular pattern once the current has been switched on. (A, ammeter; R, resistor.)

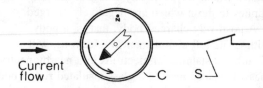

Figure 2.12. A compass is placed over a loadbearing conductor. (C, compass; S, switch.)

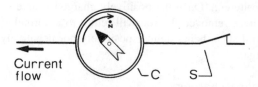

Figure 2.13. A compass is placed above a loadbearing conductor. (C, compass; S, switch.)

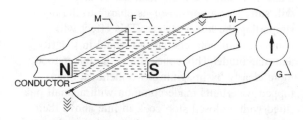

Figure 2.14. When a conductor is moved at right angles across a North–South magnetic field, a tiny electrical potential is generated within the conductor. (F, flux; G, galvanometer; M, magnet.)

The physical effect of the invisible magnetic flux on free valence electrons may be discribed as *electrical pressure* or *voltage*, whereas the uniform movement of negatively charged electrons can be described as *current flow*. This is voltage and current flow in its most basic form.

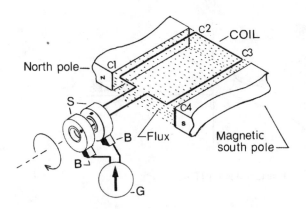

Figure 2.15. A generator is its most basic form. (B, brushes; G, galvanometer; S, slip rings.)

Simple generator

A generator is a device which converts mechanical activity into electrical energy by rotating a series of conductors through a North–South magnetic field at right angles to that field. In basic form a generator consists of a powerful permanent magnet, known as a field magnet, accompanied with an armature consisting of a single rectangular loop of lightly insulated wire. This, theoretically, would then be free to rotate around a horizontal axis routed through the centre of the magnetic field as illustrated in Figure 2.15. When the coil is rotated by a mechanical or manual primemover, the section of coil boarding the two permanent magnets (C1–C2 and C3–C4), slices through the invisible magnetic flux at right angles. The energy dispensed by the magnetic flux knocking loosely held electrons from their valency shells provides an electromotive force within the coil of wire. The magnitude of the induced e.m.f. will depend on the strength of the magnetic field and the pace at which the conductor cuts through the flux. This is maximised when the coil is horizontially positioned and it is at this point that the induced e.m.f. is at its greatest (Figure 2.16). Immediately the coil moves from the horizontal plane the e.m.f. reduces proportionally in value to zero volts. At this point the coil is vertically positioned; C1–C2 and C3–C4 are ranged *parallel* to the lines of magnetic flux (Figure 2.17).

At this juncture it is interesting to mention that virtually no e.m.f. is induced into the C1–C4 and C2–C3 portions of the coil as this section rotates in a plane which is parallel with the lines of magnetic flux. As it never crosses the magnetic flux at right angles no electromotive force is induced into the moving coil.

As the coil approaches its vertical position the induced e.m.f. rises to its maximum value and then changes direction as soon as this point of reference has been passed. This may be confirmed by *Fleming's Right Hand Rule* (see p. 151). The two free ends of the coil are connected to a pair of slip rings and generated power is routed to a fixed wiring arrangement by means of a carbon brush sprung-fixed against each of the slip rings (Figure 2.18). This is the principle of generating alternating current in its crudest form. Machines used to produce this form of electricity are known as *alternators* and are rated in kilovolt-amps (kVA).

Direct current generator

The simple alternator, schematically illustrated as Figure 2.18, may be converted into a direct current *generator* sometimes refered to as a *dynamo*, by replacing the two slip rings with a single ring split

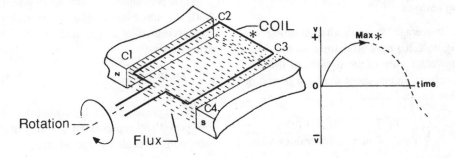

Figure 2.16. At this point the e.m.f. generated is at its greatest.

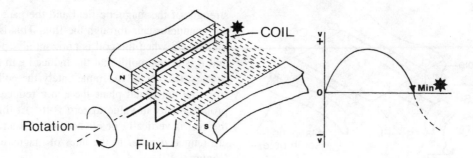

Figure 2.17. At this point the e.m.f. generated is at its lowest.

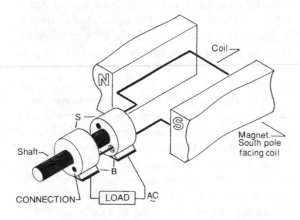

Figure 2.18. The free ends of the coils are connected to slip rings, S, and carbon brushes, B, which transfer power to the external load.

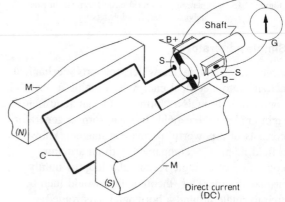

Figure 2.19. Simple dynamo to produce direct current. (C, coil; B, brushes; G, galvanometer; M, magnet; S, slip rings.)

through the middle, each half insulated from each other. Two carbon brushes placed opposite each other and sprung-clipped firmly to the newly designed commutator allows current to be drawn off and consumed as shown in Figure 2.19.

The root-mean-square (RMS) value of an alternating voltage

The mean, or average, value of an alternating voltage (Figure 2.20) may be defined as the square root of the average sum of the squares of the individual voltage values taken during a period of one complete cycle. Hence,

$$\text{RMS value} = \sqrt{\frac{V_1^2 + V_2^2 + V_3^2 + V_4^2 + \ldots}{\text{Total of number of values taken}}}$$

[2.1]

The RMS value, or effective value as it is sometimes known, is an important factor in an alternating voltage as it governs the heating effect (IR^2) produced within a conductor. It corresponds closely in value to a *stable* voltage which produces the equivalent heating effect as the alternating voltage generated.

When presented as a graphical sine wave (Figure 2.20) the maximum amplitude or peak voltage can be represented as a. The RMS value can then be expressed as:

$$\text{RMS} = \frac{a}{\sqrt{2}}$$

[2.2]

This expression may be mathematically transposed and represented in terms of a, the maximum or peak voltage. Hence,

$$a = \text{RMS} \times \sqrt{2}$$

[2.3]

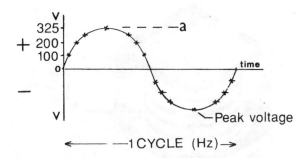

Figure 2.20. Individual voltages taken over one complete cycle in order to calculate the RMS value of the supply.

As an example, consider the following:

Calculate the maximum voltage of an alternating supply given that the RMS value is 230 volts
Referring to Expression [2.3] and substituting for known values:

$$a = 230 \times \sqrt{2}$$

$$= 230 \times 1.4142$$

$$= 325.269 \text{volts (maximum or peak voltage)}$$

All alternating current (AC) measuring instruments used in electrical engineering and installation work are calibrated to read in RMS values.

Electricity resulting from magnetism

Faraday, in 1831, carried out an experiment which led to the development and manufacture of the first commercial dynamo designed to power a marine lighthouse.

A cylindrical coil of copper wire, insulated by dry string laid by the side of the conductor, was attached to a primitive galvanometer. (This is an instrument used for measuring minute electrical currents and depends upon the electromagnetic effect produced.) *Faraday* found that when a strong permanent magnet was plummeted rapidly into the centre of the coil the galvanometer needle moved briefly to one side. Removing the magnet produced a similar but opposite effect. The needle momentarily moved in the reverse direction. *Faraday* had discovered electromagnetic induction. Figure 2.21 shows this concept in schematic form.

When a current is induced into a coil of wire by a magnetic effect the phenomenon is known as *electromagnetic induction* and is a direct result of the magnetic flux cutting through the coil at right angles to the wire. Again, the prime mover is the magnet flux forcing loosely held valence electrons from their regular orbits to collide with neighbouring valence electrons. This knock-on or 'domino' effect is current flow at its most basic form.

Induced current
The strength of an induced current is dependent upon five fundamental factors:

1. The number of turns of wire forming the coil. *An induced e.m.f. is directly proportional to the number of turns of wire.*

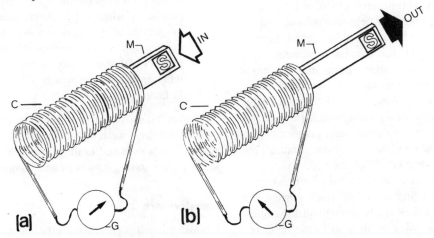

Figure 2.21. Faraday's experiment with electromagnetic induction. (C, coil; G, galvanometer; M, magnet.)

2. The material the wire is made from. *Silver or copper would be a good choice as both are excellent conductors of electricity.*
3. The temperature of the wire forming the coil. *The resistance of a conductor such as silver or copper increases with an increase in temperature.*
4. The strength of the magnetic field. *The greater the density of magnetic flux per square metre, the more effective the magnet will be.*
5. The velocity of the magnetic body entering the coil. *The e.m.f. induced is proportional to the pace of the magnet moving through the coil.*

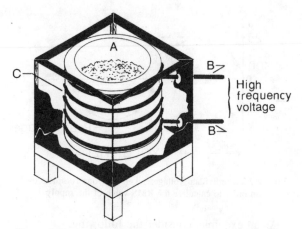

Figure 2.22. An induction furnace. (A, crucible or melting pot; B, copper pipes, watercooled; C, coil formed by copper pipes.)

Electromagnetic induction is vital to many aspects of electrical engineering and practical examples may be drawn from the following:

1. *Choke* A high induction coil of low resistance which allows direct current to flow but prohibits alternating current due to the phenomenon of self-induction.
2. *The induction furnace* This is based on high-frequency alternating current producing eddy currents of which heat is a by-product. This concept is able to generate enormous quantities of energy which is sufficient to melt metal placed in a crucible (an earthen pot for melting ores and metals). The coil is made from copper tubing in which cold water is pumped to keep it cool. The high frequency administered produces a rapidly changing magnetic field and induces eddy currents into the metal to be melted. Figure 2.22 illustrates this concept in summary.
3. *Squirrel cage motor* Currents induced into the rotor produces mechanical momentum. When a three-phase supply is applied to the stator windings a rotating magnetic field is produced and the subsequent magnetic effect induces an e.m.f. into the rotor windings by mutual induction. Since the cage windings are short circuited by means of a brass ring placed at either end of the rotor, current will be able to flow freely throughout the caged circuit. In doing so, they will develop their own magnetic fields and interact with the field generated by the stator windings.

The stator's rotating magnetic field, known as the *synchronous speed*, has the effect of dragging the rotor around with it. The current produced in the closed loop of conductors serving the rotor opposes the very current producing it. The only way to oppose this change is for the closed loop formed within the rotor to gyrate in the same direction as the rotating magnetic field generated within the windings of the stator. Figure 2.23 illustrates a typical wiring arrangement serving a squirrel cage induction motor.

4. *Transformer* An alternating current, applied to the primary winding is induced into the secondary winding by mutual induction. The magnitude of the induced e.m.f. will depend on the value of the applied voltage and the number of turns of wire in both sets of windings, details of which will be dealt with in later paragraphs.

Faraday's discoveries were soon put to practical use and dynamos were designed and built to serve marine lighthouses and in the early 1880s the first public power station was built and commissioned.

Transformers

A transformer is a simple but efficient device which is widely used throughout the electrical industry. It

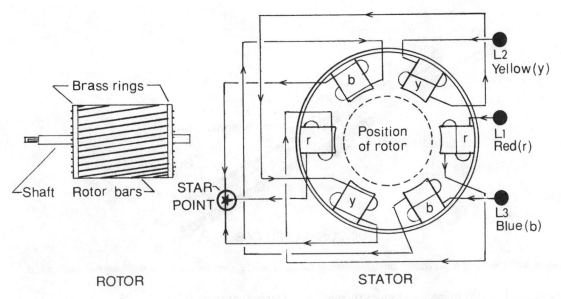

Figure 2.23. A typical wiring arrangement serving a squirrel cage induction motor.

has no moving parts and may be built to any size to cater for any assignment. It can be described as a device by which an alternating voltage is reduced or increased in value without varying the frequency of the supply.

There are three basic types which are used in industry today:

1. *Step-down transformer* Used in extra low-voltage circuits, radio and television receivers and power distribution.
2. *Step-up transformer* Used in television receivers, commercial fly-catchers and power transmission, etc.
3. *Isolating transformer* Used in television and

radio receivers and associated with TN–C earthing arrangements.

Principally, a transformer consists of a number of thinly shaped insulated plates of soft iron, approximately 0.35 mm thick, riveted together to form a laminated core. The design of the core is an important factor governing the efficiency of the device. If, for example, it was constructed from a solid block of iron, additional eddy currents would develop within the iron work causing power loss and heating problems.

A *primary coil* of low-resistance insulated wire is wound on to the laminated iron former. A *secondary* coil, physically separated from the primary

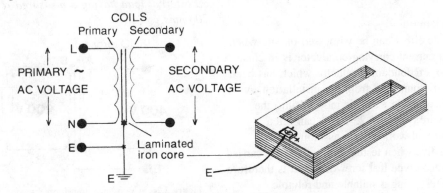

Figure 2.24. A basic transformer.

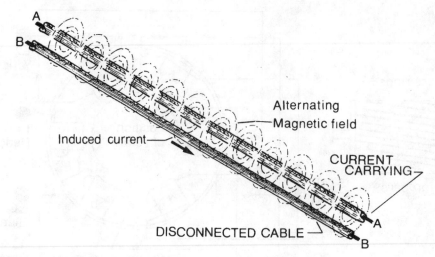

Figure 2.25. Conductor A, the current-carrying conductor, induces a small measurable e.m.f. into the abandoned conductor, B.

arrangement, is added to the iron former (Figure 2.24). Often a ratio is applied to the transformer, as for example, 1 : 5. This implies that for every one primary turn there are five secondary turns.

The working concept

When an alternating voltage is applied to the primary winding an electromagnetic field is produced which rises and falls in tune with the frequency of the supply. The physical change of magnetic flux brought about by the frequency of the supply induces current into the secondary coil. This is known as *mutual inductance*. The magnitude of the induced electromotive force depends on the voltage applied to the primary winding and the number of turns of wire in each respective coil.

A similar example

A comparable effect can be witnessed on site when an insulated current-carrying conductor is in close proximity to a redundant conductor which has been electrically disconnected from the installation. A small but measurable e.m.f. is induced into the abandoned wire (Figure 2.25). This can often mislead electrical workers into assuming that a conductor is live when testing is ill-advisedly carried out using a neon type test screwdriver. It is therefore far wiser to test using a suitable and reliable voltmeter so that mistakes may be minimised.

Current drawn

The transformer is totally dependent on an alternating magnetic flux which induces an electromotive force into the secondary winding. It appears at first that something is being gained for virtually nothing as the induced voltage can often be higher than the voltage applied to the primary winding. By calculation it can be proved that power consumed in both sets of windings is theoretically equal.

As an example, consider the following and Figure 2.26:

A 2:1 step-down transformer has a primary winding consisting of 800 turns of wire and is served with a 400 volt AC supply. The secondary winding provides for a 200 volt supply and is connected to an external load having a measured resistance of 20 ohms.

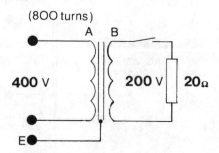

Figure 2.26. A step-down transformer serving a 20 ohm load. (A, primary winding; B, secondary winding.)

Calculate the power generated in both the primary and secondary winding.

Solution:
By applying Ohm's Law, the current drawn from the secondary winding may be calculated.

$$\text{Current} = \frac{\text{Secondary voltage}}{\text{Resistance of secondary load}} \quad [2.4]$$

Substituting for known quantities,

$$\text{Current} = \frac{200}{20}$$
$$= 10 \text{ amps}$$

Once the current in the secondary winding has been calculated, the following expression is used to find the current flowing in the primary winding.

$$\frac{V_p}{V_s} = \frac{I_s}{I_p} \quad [2.5]$$

where V_p is the primary voltage
V_s is the voltage induced into the secondary winding
I_p is the current, in amps, drawn from the primary winding
I_s is the current drawn from the secondary winding.

Substituting for known values,

$$\frac{400}{200} = \frac{10}{I_p}$$

Bringing the expression in terms of I_p by cross-multiplying and dividing each side of the equation by 400,

$$400 \times I_p = 200 \times 10$$
$$I_p = \frac{2000}{400}$$
$$= 5 \text{ amps}$$

Power generated
It seems at first sight that an additional factor has been gained but by calculating the total power generated in both primary and secondary windings it is clear that this is not the case.

Given that:

$$\text{Primary power in watts} = I_p \times V_p \quad [2.6]$$

and substituting for known quantities,

$$\text{Primary power} = 5 \times 400$$
$$= 2000 \text{ watts}$$

Given that:

$$\text{Secondary power in watts} = I_s \times V_s \quad [2.7]$$

and substituting for known values,

$$\text{Secondary power} = 10 \times 200$$
$$= 2000 \text{ watts}$$

It is clear from calculations made that equal power is generated in both the primary and secondary windings, and nothing has been gained.

Energy losses
The transformer is a very capable device, having an average efficiency of 96 per cent. It is designed so that very little energy is wasted as the following summary will help to confirm:

1. A well-designed core ensures that virtually all magnetic flux generated by the primary winding is captured by the secondary winding.
2. The core is laminated and each iron lamination is insulated from the next. Laminations help to reduce unwanted eddy currents which circulate throughout the iron work causing heating and power loss problems.
3. Both sets of windings are formed from low-resistant lightly insulated copper wire.
4. Laminations are constructed from soft iron with high magnetic qualities.

In a super-efficient transformer where losses are virtually negligible, the following expression may be accepted:

$$\text{Primary power input} = \text{Primary power output} \quad [2.8]$$

In practical terms this is far from being a reality. Power losses are responsible for wasted energy and can originate from two sources:

1. Copper losses
2. Iron losses.

Copper losses

Current flow within the primary and secondary windings produces an intrinsic power loss due to the natural resistance offered by the copper windings. This can be calculated by use of the following expression:

$$\text{Power loss in watts} = I^2 \times R \qquad [2.9]$$

where I is the current flowing in amps
  R is the resistance of the windings in ohms.

As an example, consider the following:

Calculate the copper loss in watts of a secondary winding serving a small step-down transformer given that the resistance of the secondary winding is 0.35 ohms and the connected external load is drawing 10 amps.

Referring to Expression [2.9] and substituting for known values,

$$\text{Power loss in watts} = 10^2 \times 0.35$$

$$= 100 \times 0.35$$

$$\text{Power loss} = 35 \text{ watts}$$

In practical terms it is difficult to overcome this problem but one day perhaps superconductivity will provide an answer.

Iron losses

Eddy currents are responsible for power loss within an iron-cored transformer. Magnetic flux, produced by the primary winding, is responsible for tiny current flows circulating within the laminated iron core. It is as though the iron core is acting as a supplementary secondary winding. These currents are totally unwanted and contribute towards heating problems and loss of power.

Resolving this problem can be expensive. One way is to design and build a transformer using a *dust core*. As the name suggests, this is a core made from iron dust particles mixed with an insulating compound. Another solution would be to use a high magnetic permeability *ferrite* core. Ferrite is a collective name given to a group of ceramic materials that have ferrimagnetic properties which are not susceptible to eddy currents.

Magnetic power losses

Another power loss problem is brought about from magnetic flux circulating within the iron core of the transformer. Power loss is directly proportional to the area of the theoretical magnetic hysteresis loop and the frequency of the supply. This has only been mentioned in order to provide insight into the problems that are coupled with power loss in iron-laminated transformers. Reviewing the theoretical details of the hysteresis loop falls beyond the scope of this book.

Power transmission

A brief historical background

In pioneering days, cables as we now know them did not exist. Electricity companies distributed power to homes and businesses by means of conductors formed from copper bars supported by porcelain insulators. This introductory arrangement was placed into ducting and filled with bitumen.

In 1881 the first public power cable was laid which consisted of D-shaped copper bars, insulated by means of spacers. The complete assembly was drawn into a steel conduit and filled with bitumen.

The very first electrical installations were supplied from locally sited direct current dynamos. It was not until after 1890 that alternating current, AC, was widely used when, at *Deptford*, the very first power station to transmit 10 000 volts, at 60 Hz was commissioned. In those days a high-voltage cable crudely consisted of an approximately 23 mm diameter copper tube covered in brown paper impregnated with a mineral wax. This was then placed into another copper tube insulated with impregnated paper. The completed assemblage was then drawn into a cast-iron tube. Although crude compared with today's technology, it lasted many years and was finally decommissioned in the early 1930s.

In the town of Godalming, England, electricity

was available to the general public as far back as 1881 where it was distributed by means of crude conductors placed in guttering. The supply voltage was about 100 volts so electrical engineers at the time decided it was not economical to extend the radius of their transmission lines any further than 1 mile (1.6 km) from the generator. The power loss within the conductors would have been enormous and the only way to conquer this difficulty was either to increase the size of their supply cables or to increase the value of the generated voltage. For this reason it was judged to be uneconomical to expand the service lines.

By 1936 the National Grid System had been established to serve a major part of the country. A triple-phase high-voltage system using stranded aluminium conductors, often supported by an internal steel core to provide mechanical strength, was implemented throughout the land, bringing electricity to rural areas where once oil lamps, solid fuel ranges and secondary cell-powered radios were commonplace.

Some of the tallest pylons built, 148.5 metres high, are sited at *Dagenham* to provide safe passage for grid voltage distribution cables across the *River Thames*. The three current-carrying conductors were accompanied by a solitary earth wire ranged and clamped to the top section of the pylon which also acted as a lightning conductor.

The effects of the introduction of electricity
Gradually electricity became accessible to the majority of the population throughout the country. Average families were primarily interested in using this new source of power to provide light in their homes. Lighting was previously supplied by oil lamps and candles in rural areas, although townsfolk sometimes had access to gas-lit lamps. So imagine the magical impact that was made upon people by the simple act of flicking a switch in order to illuminate a room.

Gradually over the following years modest electrical appliances were purchased. Electric irons replaced flat irons which had to be heated on an oven hob. Low-wattage kettles, mains radio receivers, table lights and electric fires were among sought-after items.

The introduction of pioneering washing machines made clothes washing a much less arduous task. Refrigerators enabled perishable goods such as milk, cheese and meat to be safely stored much longer than in the previously used larder. This meant that people were far less likely to run the risk of eating contaminated food and becoming ill. When freezers entered the marketplace a whole new world of frozen foods was opened up to the general public. Food shopping did not have to be an everyday task and surplus garden produce could be safely frozen and kept in the home freezer.

All these changes meant that the quality of life for most people was significantly better thanks to improved health and more leisure time.

In recent years we have seen numerous developments advancing both selection and choice of electrical appliances and devices beyond our imagination. Wide screen television, video decks and compact disk players were all developed to provide entertainment within the home. Computers which store, retrieve and transmit data were designed to meet the requirements of commerce and industry.

Passive infrared detectors and residual current devices have been fashioned to provide a degree of security and protection throughout our daily lives.

The Nation Grid System has fundamentally transformed the way we live; the quality of life has changed beyond expectation for the average person. With regret we have quickly learned to take electricity for granted. It is only when the distribution network is disrupted by gale-force winds or thunderstorms that we realise just how important this commodity really is. Then we wonder how our forefathers used to live without the use of electricity!

Modern systems
Today alternating voltages are generated and fed into the National Grid supplying homes and businesses throughout the country (Figure 2.27).

There are many different forms of energy from which electricity can be generated. Heat energy, obtained through the combustion of fossil fuels, is generally used to produce steam to turn giant turbines providing mechanical means to motivate the alternators. However, other prime movers are also employed which, in listed form, include the

Figure 2.27. An electricity pylon serving the National Grid.

following:

1. Gas
2. Hydro
3. Nuclear
4. Oil
5. Wave motion (experimental)

Most of our electricity is generated between 11 000 and 33 000 volts by a machine known as an *alternator* driven at high speed by a steam turbine. The generated electricity is transformed in value to 132 000, 275 000 or 400 000 volts and fed into the National Grid (Figure 2.28).

The grid system was first conceived in the 1920s. Since then it has been altered and modified into an interconnecting network of power lines and now provides an efficient fail-safe means for high-voltage transmission throughout the country. Should a major line fault or mechanical breakdown develop at any contributing power station, current can

immediately be drawn from other transmission centres along alternative lines. This technique provides an efficient, reliable energy distribution system (Figure 2.29).

Underground developments: high-voltage transmission

Although environmentally far more acceptable, high-voltage cables are not generally placed underground for the following reasons:

1. Cost
2. Insulation considerations
3. Heating problems

1. *Cost* The average cost involved when installing an underground high-voltage cable is a staggering twenty to thirty times greater than installing overhead lines supported by steel pylons. Many cables are often required in this type of installation and when laid adjacent to each other can occupy a width of some 12 metres.

2. *Insulation* The unit cost of insulation conductors operating at very high voltages is extremely expensive. Dry air is an excellent insulator when the cables are routed overhead and of course it is free! This alone is a large financial saving for the power

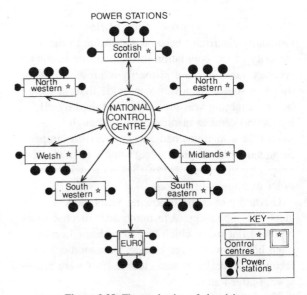

Figure 2.28. The production of electricity.

unfortunately, results in extensive areas of top soil being dried out, producing a barren and virtually sterile area where difficulty is experienced in cultivating any form of vegetation that will withstand and flourish in the dry conditions forced upon it.

Distribution of electricity to the consumer

Electricity, generated at voltages between 11 000 and 33 000 volts, is transformed to a higher voltage and distributed by way of the National Grid system. Substations, built in principal locations, provide means by which power can be taken from the National Grid and reduced to 11 000 volts.

In rural areas distribution is by means of uninsulated conductors carried on high-quality insulators attached to wooden poles impregnated with a preservative (Figure 2.30). These are routed from local substations supplying energy to community transformers cradled safely out of reach between two sizeable wooden poles. At this point the voltage is reduced again, but to a conventional 400 volts (Figure 2.31). Power is then distributed by a network of overhead insulated phase conductors carried by insulators on wooden poles as illustrated in Figure 2.32 and provides means of dispensing both single- and triple-phase power to the consumer.

Aesthetic considerations
Local electricity companies strive to blend in pole transformers with the environment. Often they are painted leaf green or grey and sited in a background of trees. In towns and cities substations are harmonised with their surroundings disclosing just a mere hint of their true role by displaying a 'DANGER HIGH VOLTAGE' warning notice fitted to the door. Power is then distributed to the consumer by means of a network of underground cables (Figure 2.33).

The advantage of high-voltage power transmission

In order to reduce heating and power loss caused by the natural resistance of a conductor, a very heavy cable would have to be used to transmit

Figure 2.29. A smaller pylon, carrying 132 kV, serving the National Grid.

transmission company which undertakes the design task for major projects but is restricted by budget rationing.
3. *Heat* High-voltage conductors consume vast amounts of power due to the $I^2 \times R$ loss incurred within the conductor and the connected load. Large power demands produce a proportional heating problem within an underground cable. This is overcome at cost by installing a network of water pipes alongside the high-voltage cables. Pumping and cooling stations are built every third kilometre or so in order to control and monitor the temperature of the cables. This,

Figure 2.30. Wooden poles distributing power in rural areas. This type of pole carries 11 kV.

Figure 2.31. Rural pole transformer serving a small community.

power over long distances. This is because power loss in a conductor is equal to the sum of the current squared and the internal resistance of the conductor. Hence:

$$\text{Power loss} = I^2 \times R \qquad [2.10]$$

where I is the current flowing in the conductor in amps, and

R is the internal resistance of the conductor in ohms.

However, there is an alternative method. When electricity is transmitted at high voltages a power loss of far smaller proportions occurs and this can be demonstrated by the following example.

Power totaling 20 kW (20 000 watts) is transmitted by means of an aluminium conductor having a measured internal resistance of 0.7 ohms.

Calculate the internal power loss within the conductor when the voltage applied is (a) 230 volts and (b) 275 000 volts.

Solution:
(a) Applying a voltage of 230 V.

$$\text{Current consumed in amps} = \frac{\text{Wattage } (W)}{\text{Voltage } (V)}$$

$$[2.11]$$

Substituting for known values,

$$\text{Current consumed} = \frac{20\,000}{230}$$

$$= 86.9 \text{ amps}$$

Power loss in a conductor $= I^2 \times R$ (Expression [2.10])

$$\text{Power loss} = 86.9^2 \times 0.7$$

$$= 5286.12 \text{ watts (5.286 kW)}$$

(b) Applying a voltage of 275 000 V.

$$\text{Current consumed in amps} = \frac{\text{Wattage}}{\text{Voltage}}$$

(Expression [2.11])

Substituting for known values,

$$\text{Current consumed} = \frac{20\,000}{275\,000}$$

$$= 0.072 \text{ amps}$$

Power loss in a conductor $= I^2 \times R$ (Expression [2.10])

$$\text{Power loss} = 0.072^2 \times 0.7$$

$$= 0.0036 \text{ watts}$$

Compared with the high-voltage example, the energy wasted at 230 volts resulting from the internal resistance of the transmission conductor is unacceptably excessive. Due to this, power is

Figure 2.32. Street poles distributing electricity to homes and businesses in rural areas.

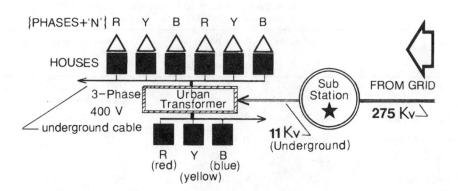

Figure 2.33. In towns and cities electricity is distributed by means of a network of underground cables.

transmitted at an elevated voltage throughout the National Grid. Alternating current is used as it can be efficiently converted to a higher voltage with a minimum of loss. If a direct current system were to be applied it would be virtually impossible to transmit power efficiently and cheaply over long distances.

Types of power stations

Table 2.2 examines a small selection of British power stations, their kilowatt output together with the type of fuel used to motivate their alternators.

Control of pollution

With the demand for power increasing every decade, the control of pollution is paramount to the environment.

The natural residue from coal-fired power stations is used for a variety of commercial purposes. It is added to lightweight building blocks during the production stages, sold to manufacturers of cement products and widely used in the aerospace industry. Building and land reclamation projects are sold huge amounts of pulverised ash as a means of providing an infill when ground levelling schemes are implemented.

Some types of nuclear fuel, when spent, are safely removed for reprocessing then used again. Other forms of expended high-level radioactive fuel are carefully removed and placed in stainless steel containers which are externally lined with concrete and buried at great depths on land or under the sea bed.

Siting of power stations and transmission lines: aesthetic considerations

At the planning stage considerable time and thought is spent deciding where a power station will be sited. They are usually built well away from urban areas and placed in rural backgrounds where conditions are favourable, for example, near rivers for cooling purposes, major roads for access and coal mines for fuel.

In order to harmonise with the environment, landscaping projects are commissioned to provide an assortment of indigenous trees and shrubs. Where woodlands have been acquired resulting from the acquisition of building land, field study centres of nature trails have been established providing access for school children to study and assess the ecological changes brought about by a synthetic and altered environment.

It is well known that many different species of fresh water fish and eels thrive in the warm waters produced as a by-product from power stations. This has given rise to the development of major fish-farming projects at selected sites throughout the British Isles. At *Cynrig* near *Brecon*, over 10 000 young salmon are reared every year to be returned to the *River Usk*. This more than compensates for immature fish which are killed resulting from being drawn into the power station's cooling system.

TABLE 2.2   A selection of British power stations

A kilowatt is defined as a measurement of power equal to 1000 watts whereas a megawatt is a unit of power equal to 1 000 000 watts

Site	Prime mover	Power output in kilowatts (kW) or megawatts (MW)	Commissioning date
Berkeley	Nuclear	4 × 83 MW	1962: the world's first nuclear power station
Carmarthen Bay	Wind	200 kW	November 1982
Chagford, Dartmoor	Hydro	31 kW	1940
Cowes	Gas	110 MW	1982
East Yelland	Coal	6 × 31 MW	1953
Fawley	Oil	4 × 500 MW	1971

Transmission lines

Pylons and their associated transmission lines are carefully sited to follow the environmental profile of the countryside but final judgement is usually the responsibility of the local planning authority. However, this policy does not seem to reflect decisions made as to where 11 000 volt wooden pole transmission lines are placed, as Figure 2.34 will illustrate.

Sometimes transmission lines have to be placed underground where environmentally they would be more acceptable and not contribute to the depletion of the countryside. Where this is not possible for technical reasons, for example, a route through the base of a mountain, transmission lines are steered around sensitive rural areas in order to preserve the natural beauty of the countryside.

Electromagnetic effect generated from transmission lines

It has, for many years, been suggested that living under or adjacent to high-voltage transmission lines can produce ill effects. In listed summary, these are alleged to include the following:

1. General unease
2. Stress-related illnesses
3. Headaches
4. Skin rashes
5. Cancer of the blood (leukaemia)
6. Depression
7. Heart palpitations
8. Urogenital problems
9. Watery eyes

It has recently been put forward that harmful side-effects associated with 400 000 volt transmission lines are generated by one or a combination of the following factors:

1. Strong electromagnetic fields, *These have been measured and found to be in excess of 1000 micro-gauss in certain areas.*
2. Air molecule ionisation. *The mean density of ionised molecules per cubic metre is dependent on the strength of the magnetic field, the surrounding terrain and the speed and direction of the ground level winds.*
3. Low-frequency sound waves. *These are*

Figure 2.34. Transmission lines serving rural areas.

generated by the wind blowing across the transmission lines and through the pylon's infrastructure.

It is obvious that additional meaningful research has to be carried out by an independent body in order to establish the truth behind these allegations. In the meantime it might be prudent not to set up home under or near to a 400 kV transmission line!

Summary

1. *William Gilbert* (1540–1603) coined the term 'electricity' from the greek word *elektron*, meaning amber.

2. An atom is the smallest fragment of an element that is capable of engaging in chemical reaction. It can be subdivided into three parts: *proton*, a positive particle; *neutron*, the neutral particle; and *electron*, the negative particle.

3. Like charges repel one another. Unlike charges attract.

4. Silver is a better conductor of electricity than copper.

5. Materials which have many electrons in their outer shell are known as insulators and are bad conductors of electricity.

6. Current flow is a physical movement of negatively charged subatomic particles called electrons.

7. *Oersted* discovered that electricity and magnetism are related phenomena.

8. A generator converts mechanical movement into electrical energy by rotating a series of conductors through a *North–South* magnetic field at right angles to that field.

9. The average value of an alternating voltage may be defined as the square root of the average sum of the squares of the individual voltage values taken during a period of one complete cycle.

10. *Faraday*, in 1831, carried out an electrical experiment which led the way to the development of the first commercial dynamo.

11. Electromagnetic induction is important to many aspects of electrical engineering. Examples are drawn from the following: the common choke, the induction furnace, a squirrel cage motor and the transformer.

12. There are three basic types of transformer:
    (a) Step-down; used for extra low voltage circuits
    (b) Step-up; used for power transmission
    (c) Isolating transformer; used in association with TN–C earthing arrangements.

13. A transformer is totally dependent on an alternating magnetic flux which induces an electromotive force into the secondary windings. Both sets of windings are isolated from each other.

14. Eddy currents are responsible for power loss within an iron-cored transformer. A dust-cored transformer will virtually resolve the problem.

15. Alternating current was not commercially used until 1890.

16. There are many different forms of energy from which electricity can be generated. Examples are: coal, gas, hydro, nuclear, oil, wave motion and wind.

17. Electricity is generated between 11 000 and 33 000 volts. It is then transformed in value to 132 000, 275 000 or 400 000 volts and fed into the National Grid.

18. The National Grid System was established to serve a major part of the country by 1936. Stranded aluminium conductors were used, often supported by an internal steel core to provide mechanical strength.

19. Early consumers of electricity were eager to receive lighting in their homes. Gradually over the years modest electrical appliances were purchased.

20. Underground high-voltage transmission cables, although environmentally more acceptable, are generally not placed beneath the soil for the following reasons:
    (a) Cost
    (b) Insulation considerations
    (c) Heating problems

21. Substations provide means by which power taken from the National Grid can be reduced to 11 000 volts. The converted voltage is routed across country to local community transformers where it is reduced to a familar 400 volts to provide power for domestic, commercial and industrial consumption.

22. When electricity is transmitted at high voltages the power loss, due to the natural resistance of the conductor, is far smaller than if domestic voltages were to be used.

23. Some types of nuclear fuel, when spent, can be safely removed for reprocessing then used again. Unwanted high-level radioactive fuel is carefully removed and placed in stainless steel containers externally lined with concrete and buried at great depths.

24. Power stations are built away from urban areas and are landscaped to harmonise with environmental conditions. Fish-farming projects are based at selected sites taking advantage of the warm waters produced as a by-product

from the power station.

25. It has been suggested that living under or adjacent to high-voltage transmission lines can produce ill effects from strong electromagnetic fields, air molecule ionisation and low-frequency sound waves.

Review questions

1. List the three principal parts of an atom.
2. How many valence electrons are present in the outermost shell of an atom of mercury (Hg)?
3. Place the following in order of conductivity: copper, silver, quartz, mercury.
4. Confirm the following statements:
   (a) Current flow is a physical movement of positive electrons. TRUE/FALSE
   (b) Current flow is from *positive* to *negative*. TRUE/FALSE
   (c) The nucleus of an atom of copper has no roll to play in current flow. TRUE/FALSE
5. Define the root-mean-square (RMS) value of an alternating current.
6. Briefly describe the principle of an electric choke.
7. List three basic types of common transformer.
8. Suggest two reasons why energy is wasted in a iron-laminated transformer.
9. How were early current-carrying conductors formed and what method was adopted to insulate them?
10. Name three prime movers used in the generation of electricity.
11. What considerations should be addressed when designing an underground high-voltage transmission cable installation?
12. Selected power stations transform their output of electricity generated and supply the National Grid at the following voltage:
    (a) 11 000 V
    (b) 270 000 V
    (c) 400 000 V
    (d) 450 000 V
13. State the advantage of high-voltage power transmission.
14. What are the reasons for the alleged harmful side-effects attributed to high-voltage transmission lines?
15. List two commercial uses for the natural residue from coal-fired power stations.

Handy hints

1. The term *SELV* was originally formed using the initials of the definition, *Safety Extra Low Voltage*.
   It is now agreed that reference should not be made to the meaning of the individual letters as no voltage should be regarded as safe, however low.

2. Never rely on stranded wire armouring as a means of providing an earth. Often there is insufficient cross-sectional area to satisfy the demands of the *Wiring Regulations*. See Regulation 543–02. In outside situations corrosion can occur on semi-exposed steel armouring. Corrosion can also occur when copper and aluminium conductors are mechanically bonded together.

3. To convert *horsepower* into *kilowatts*, divide the horsepower rating by 1.341 or multiply the total horsepower by 0.746.

4. Changing a star-delta contactor coil is made far easier by labelling all conductors and integrally fitted wires. This way mistakes are minimised.

5. A locally burnt wire, originating from a protective device, is probably caused by a badly fitted or loose conductor. Should a fault such as this be discovered, a check should be made on all connections serving adjacent protective devices. This will ensure that all conductors have been terminated correctly.

6. Avoid targeting powerful spotlights too close to material objects. Allow at least 0.5 metres to minimise the risk of fire.

7. Visually inspect and instrumentally test portable generators on a regular basis to confirm safety in use. Results should be recorded formally in a log. The following checks should be carried out:
   (a) Condition of the sockets.
   (b) Tightness of all serviceable connections.
   (c) Operational checks on all controls and functional switching.
   (d) Condition of all protective enclosures and mechanical parts.

(e) Insulation resistance tested from the generator windings to the frame.

(f) Earth continuity tested from the earthing terminal of each service socket to the common earthing point attached to the frame.

(g) Voltage tested.

(h) Load tested.

(i) Functional protective over-current device tested.

(j) Fuel and oil checked.

3 Fire detection and alarm systems

In this chapter: Fire alarms: early and modern systems; wiring techniques. Fire detection and alarm accessories. Zoning. Monitoring an installation. Testing, commissioning and keeping records. Intruder alarms; early and modern systems. Tamper circuits, intruders and alarm accessories; wiring techniques and testing. Emergency lighting. Regulations governing installation requirements, Types of emergency lighting. Positioning. Testing and health and safety. Nurse call system.

A correctly installed fire alarm installation is of paramount importance and can be compared to any other electrical undertaking. Life could be lost and property damaged resulting from carelessly or incorrectly connected fire detection and alarm equipment. It is essential that the installation is carried out complying with the requirements of BS 5839, Part 1; 1988, the *Wiring Regulations* and the manufacturer's instructions.

Fire detection and alarm circuits are deemed as *Category 3* circuits and therefore must be segregated from each other and other cables sharing the installation as required by Regulation 528–01. However, when wiring is carried out using mineral insulated copper cable this demand is dropped.

In order to comply with the wiring regulations a dedicated circuit, illustrated as Figure 3.1, must be installed to supply mains power to the fire alarm control panel. In practice, wiring would be terminated in an unswitched fuse connection unit to BS 1362; the size of fuse would depend on the manufacturer's recommendation.

Previous systems

Mains-operated alarm bells controlled by manually operated call points were once a popular method of providing fire protection for architecturally designed constructions. This type of installation, often carried out using solid drawn steel conduit served with vulcanised india rubber insulated conductors, can still be seen today in premises which have never kept up with the technical advancements in the field of automatic fire detection.

A later method was provided with simple wiring arrangements comprising a trickle charger, 24 volt battery and relay unit connected to a series of manually operated call points serving an extra-low-voltage fire alarm bell. A typical installation would

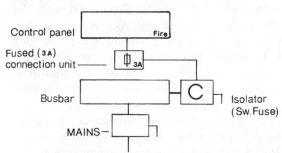

Figure 3.1. A typical dedicated circuit arrangement serving a fire detection and alarm panel. The main fire alarm switch, C, should be painted red and labelled: DO NOT SWITCH OFF.

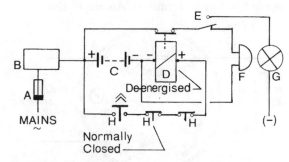

Figure 3.2. A very early type of fire alarm installation. (A, over-current protection; B, charger; C, battery; D, energised relay; E, bell silenced switch; F, bell; G, bell silenced lamp; H, call points.)

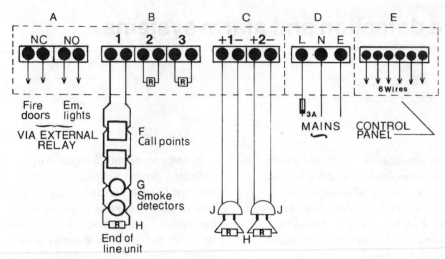

Figure 3.3. A typical modern fire alarm control panel. (A, auxiliary contacts; B, fire detection zones; C, sounder circuits; D, low-voltage mains; E, connection to remote indicator panel; F, call points; G, smoke detector; H, end-of-line resistor; J, sounders; N/C, normally closed relay contacts; N/O, normally open relay contacts.)

have been wired in a similar fashion to the schematic line diagram as shown in Figure 3.2.

Modern system: an introduction

Today the regulations and requirements governing the installation of fire detection and alarm systems are much stricter and must comply with the demands of BS 5839, Part 1: 1988. It is wise to obtain a copy of this document entitled, *Code of Practice for Systems, Design Installation and Servicing* as the *Wiring Regulations* will not apply to this type of installation.

Figure 3.3 illustrates a typical fire detection and warning system. Manual break-glass call points, smoke detectors and heat detectors are wired together in parallel formation. An end-of-line

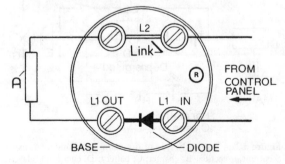

Figure 3.4. An end-of-line resistor or capacitor, A, is fitted across the terminals serving the last detector.

resistor, generally valued between 46 000 and 50 000 ohms, is fitted across the terminals serving the last detector in order to monitor the zone against open or short circuit conditions. This is illustrated in schematic form as Figure 3.4 and will be evaluated later in this chapter. Some manufacturers prefer to incorporate a capacitor but this is a drawing board decision made in the early design stages of the system. A break or short circuit occurring in any conductor would trigger the monitoring circuit to advise of a fault condition by use of audiovisual indicators on the control panel.

Circuit-serving fire detection and manual call points are known as *zones*. Generally up to 30 in number, current-consuming detectors can be supplied from one zone and as many as 120 zones can be incorporated in one fire alarm panel. The number of zones would be proportional to the floor area of the building served. As break-glass call points are not current consuming, there is no limit to the number that may serve one zone. It is important that both detectors and call points to be used in any fire detection and alarm system are compatible with the control panel. It may be financially tempting to use unmatched equipment but it is far wiser to purchase a complete system from a well-established manufacturer such as *Cerberus*.

Audio alarms are known as *sounders* and these,

together with the various types of heat and smoke detectors, will be dealt with in future paragraphs.

Wiring techniques for fire alarms

Fire detectors and sounders must be wired in a continuous parallel formation arrangement as illustrated in Figure 3.5. No spurs are permitted as this would prove difficult to monitor for breaks and short circuits which might occur. Monitoring will be dealt with in full later in this chapter.

Circuits from the control panel should be as short as possible as many manufacturers require a maximum of $100\,\Omega$ per fire detection *loop*, as outlined in Figure 3.6. In practice this will not cause undue problems as the nominal impedance of a single loop of $1.5\,\text{mm}^2$ two-core cable lies between 19 and $22\,\Omega$ per kilometre.

To meet the demands of Regulation 528–01–03 to 04, also BS 5266 and BS 5839, cables serving fire detection and alarm systems must be segregated from other installations. However, separation is not required when the fire alarm system is installed using mineral insulated or *Pirelli FP 200* cable manufactured to British Standards 7629. Regulation 528–01–06 confirms.

It is wise to leave the wiring of the fire alarm system until most of the construction work on the architectural installation has been completed. This will avoid accidental damage occurring to the cables.

Keep the control panel in the packing carton and only remove when building work has been completed in the area where it is to be mounted, thus avoiding possible contamination to the unit.

What type of cable may be used?
Wiring may be carried out using either high-performance mineral insulated cable to BS 5839 and BS 6387 or a suitable fire-resistant and flame-retardant Lower Smoke Zero Halogen cable such as Pirelli FP 200.

When Pirelli FP 200 is used, leave sheath removal until the second fix stage of the installation to avoid possible damage to the insulation.

Once wired, position the terminations in a manner which will prevent other tradespeople unwittingly causing harm to the conductors.

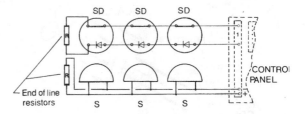

Figure 3.5. Wiring must be continuous. No spurs are permitted. (SD, smoke detector; S, sounders.)

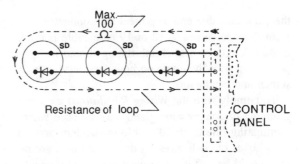

Figure 3.6. A maximum of 100 ohms per fire detection loop. (SD, smoke detectors.)

Removing the sheath of Pirelli FP 200 cable
The outer sheath can be removed by scoring the periphery of the cable with a sharp knife to the face of the aluminium screening, then flexing the cable until the workpiece snaps. The short section of snapped sheathing can then be withdrawn by twisting the cable while gently pulling and following the natural curl of the conductors. Never attempt to tug off as damage could result in the form of torn or split insulation.

For longer tails it is advisable to secure the assistance of another person. Once the sheath has been scored and ruptured, one person should firmly hold the cable a little way past the breaking point while the other carefully straightens the sheathing to be removed. With one hand placed at either end of the sheath, it may be removed by gently pulling and following the natural twist of the conductors. It is not advisable to remove small bits of sheathing at a time as the silicone rubber insulation exposed beforehand could be damaged through carelessness. It is far better to carry out the task patiently as recommended by the manufacturer of the cable.

Once the sheath has been removed, a nylon ferrule (Figure 3.7) must be slid onto the face of

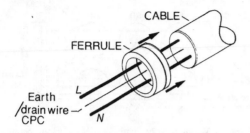

Figure 3.7. A nylon ferrule is slid onto the face of the exposed cable end.

the exposed cable end to protect the conductors from damage. When 7, 12 or 19 core cable is used, a small length of heat-shrink tubing should be positioned over the common boundary serving both screen and conductors.

To comply with the *Wiring Regulations* a grommet or compression gland must be used when terminating Pirelli FP 200 cable in standard pressed steel boxes or enclosures. In dry situations a grommet would be sufficient to meet these requirements, whereas a compression gland would be far more suitable for an installation designed for use outside.

Earthing

As a bare tinned copper protective conductor is constantly touching the sides of the aluminium screening throughout the length of the cable, there is no need to connect the aluminium sheath independently to earth. Once exposed, the bare

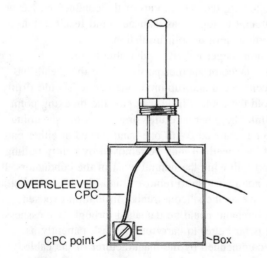

Figure 3.8. The bare protective conductor is sleeved with green–yellow oversleeving and terminated in the earthing lug provided.

protective conductor should be sleeved using green/yellow PVC oversleeving and terminated in the earthing lug provided within the accessory (Figure 3.8).

Some multi-core cables are fashioned with a small *drain wire* having a short circuit current capacity of approximately 75 amps for one second. Should a full-sized current protective conductor be required, one of the conductors must be sacrificed to serve as a full-sized protective conductor. The status of this conductor must then be changed by placing a green/yellow oversleeve throughout its entire exposed length, at both ends of the cable.

The insulation

The insulation serving the cores is made from silicone rubber which, under conditions of extreme heat or fire, is chemically converted into silicone dioxide. In this situation, silicone dioxide becomes a first-class insulator. The sheath is constructed from aluminium and covered with a hard grade PVC, laminate. Alternatively a *Lower Smoke Zero Halogen* cable can be used where strict control of corrosive gas and smoke from burning cables is demanded.

Fixing techniques

Fixing is not required when fire alarm cables are housed in inaccessible positions as long as the length of unsupported cable does not go beyond 5 metres and is placed on a dry smooth surface which would be sustained in the event of a fire. An example for this requirement would be a metal roofing purlin or a reinforced steel joist. Where fixings are required they must be made from a material which is fire retardant and spaced so that the total weight of the cable is safely supported throughout its length. The radius of any right-angled bend must be not less than six times the diameter of the cable as shown in Figure 3.9. A tight radius might look stylish but it could also be responsible for hidden internal damage to the silicone rubber insulation. This would inevitably lead to unexpected fault conditions occurring within the system.

Mineral insulated cable can also be used for fire detection and alarm installations. Details of cable management may be found in Chapter 11 of the

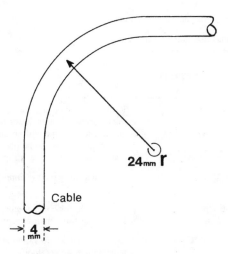

Figure 3.9. The radius of the right-angled bend must not be less than six times the diameter of the cable.

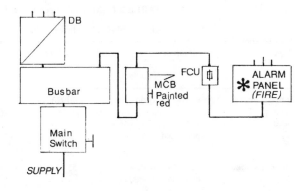

Figure 3.10. A fire detection and alarm circuit must be provided with a dedicated supply.

second edition of *Electrical Installations* (C. Shelton), published by Longman Higher Education.

Supplying the control panel with mains electricity

A dedicated circuit must always be wired to serve a fire alarm system. Ideally this should originate from a lone miniature circuit breaker supplied directly from the main busbar system and be terminated adjacent to the control panel with an unswitched 3 amp fused connection unit (Figure 3.10). The cicuit should be labelled. **Fire alarm – do not switch off**, and the accessories painted red.

Fire detection and alarm accessories

Table 3.1 profiles the manually operated break-glass call point and four of the most common automatic devices employed for fire detection.

All contributory detectors and accessories may be mixed and connected to the same zone. However, when incorporating optical smoke detectors it is important to fit the low-current-consuming type in order not to place an excessive load on the control panel relay terminals. If in doubt, seek advice as to the maximum load that may be drawn from any given zone. Specifications vary from manufacturer to manufacturer.

Other devices which may be incorporated in the design of a fire detection and alarm installation include:

- Beam detector (Gallium arsenide infrared light type)
- Duct detector (Photo-cell obscuration type)
- High-temperature heat detector (operational at 90 °C)
- Special environment call points (outdoor and industrial types)

Fire detector: ionisation principle

The automatic ionisation fire detector, once known as a smoke detector, was first marketed in 1941 by the Cerberus Company in Bad Ragaz, Switzerland. It was later completely redeveloped by Dr Ernst Meili (b. 1913–) from experimental work carried out a decade earlier by the Swiss physicist *Walter Jaeger*.

This highly developed device (Figure 3.11) functions on the principle of comparing two integrally fitted ionisation chambers of which one is free to accept minute amounts of visible and invisible particles of combustion while the other is hermetically sealed. Each chamber is served with an *extra low direct current voltage source* from a cathode and anode plate. Air molecules, free to enter the open sample chamber, become ionised by the radioactive transuranic element *americium 241*. (A transuranic element is an element which has a higher atomic number than uranium.) This gives rise to the surrounding ambient molecules becoming either positively or negatively charged. Charged molecules, suspended between the anode and cathode, are attracted to an electrode whose electrical polarity is opposite to that of the charge

TABLE 3.1   Manually operated and automatic fire detection equipment

The symbol μA indicates microamps whereas mA indicates milliamps

Type	Situation	Current consumed	Mean ambient operational temperature range (°C)	Comment
Ionisation principle (smoke head)	Corridors, bedrooms, refuge rooms and offices where smoking is not permitted	50 μA	−30 to +80 °C	Radioactive (typical value of 0.9 microcuries). Unsuitable for lounges and kitchens. Some will not tolerate high humidity. Not to be placed near to air conditioning grills
Optical smoke detector	Corridors, bedrooms, computer rooms, cellars	55 to 110 μA	0 to +70 °C	Unsuitable for garages (fumes)
Heat detector (rate of rise)	Bedrooms, cupboards	45 μA	−10 to +90 °C	Not suitable for boiler rooms, kitchens or dirty environments
Heat detector (fixed temperature type)	Bedrooms, boiler houses, kitchens, drying rooms	45 to 90 μA	−10 to +90 °C	Ideally suited where variation in temperature are common. Unsuitable for damp areas
Break-glass call point	Adjacent to exit and escape routes	None	up to +90 °C	Manually operated. Single pole operation
Fire alarm bell	Corridors, public areas, basements	24 to 28 mA	0 to +80 °C	Can be weatherproofed. Bells are fitted with a series diode to enable monitoring. Sound level, 92–96 decibels
Fire alarm siren	Corridors, public areas, outside situations	15 to 19 mA	−30 to +80 °C	Generally weatherproof. Often fitted with a volume control. Sound level 100–105 decibels

carried by the molecule (*remember*: like charges repel while unlike attract). This permits a tiny current of less than one thousand millionth of an amp to flow between the anode and cathode plate (Figure 3.12).

Figure 3.11. The automatic ionisation fire detector. (Reproduced by kind permission of *Photain Control Limited*.)

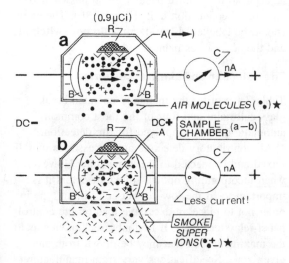

Figure 3.12. The principle of the ionisation fire detector: (a) non-fire mode; (b) fire alarm mode. (A, electron flow; B, charged plates; C, galvanometer (illustrative only); R, radioactive source '0.9 microcuries'.)

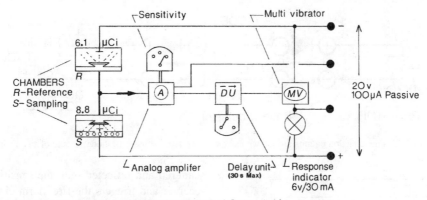

Based on the *F600* developed by CERBERUS of Switzerland

Figure 3.13. Minute change in current flow is interchanged into measurable variation in voltage.

Smoke and other products of combustion entering the chamber attract charged air molecules which envelope the invading pollution to form what is known as *super ions*. This reduces and impairs the mobility of the ions causing a drop in current to occur.

A second, but airtight, chamber connected in series formation with the sampling chamber allows minute changes in current flow to be interchanged to measurable variation in voltage by means of an analog amplication unit (Figure 3.13).

Once a predetermined response threshold has been reached, the detector transmits a signal from the amplifier to a delay mechanism which in turn communicates with the fire alarm control panel. The accepted signal automatically triggers the alarm sequence and activates the sounders. When triggered, the detector will latch in electronically until manually reset by way of the fire alarm control panel once smoke and other products of combustion have been evacuated from the sampling chamber.

The F600 *smoke detector*, developed by the Swiss manufacturer Cerberus, has incorporated options allowing three levels of response sensitivity selected by means of a 'screwdriver' switch at the base of the detector. Choice may be implemented where, for example, there are low ceilings in an area where early response is paramount. As an additional option; by placing a special blue plastic plug in a purpose-made opening provided, the alarm sequence can be electronically delayed for up to 30 seconds. This way the detector can efficiently

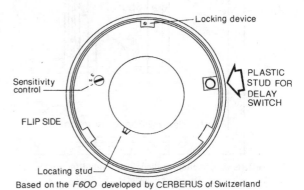

Based on the *F600* developed by CERBERUS of Switzerland

Figure 3.14. By placing a special blue plastic plug in the purpose-made opening, the alarm sequence can be delayed.

deal with fast-changing ambient conditions such as dust pollution or transient clouds of tobacco smoke. Figure 3.14 illustrates the positioning of these two important features.

Zoning

Fire detection circuits are known as *zones* and are wired in one continuous parallel circuit terminating at an end of line resistor or capacitor to enable the system to be constantly monitored. This is schematically illustrated as Figure 3.15. Spurs must not be included. It would be impossible to supervise such an arrangement. However, a light-emitting diode (LED) unit may be spurred from a detector base to advise should a remotely sited detector be triggered as described in Figure 3.16.

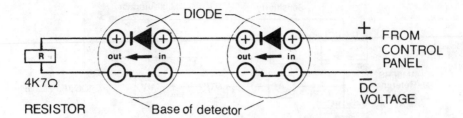

Figure 3.15. Fire detection circuits are known as zones. (A, detector base; B, diode; C, end-of-line resistor.)

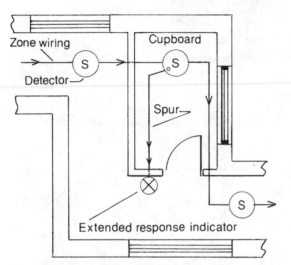

Figure 3.16. A light-emitting diode unit spurred for a remotely sited detector.

Smoke detector: optical type

The optical smoke detector, sometimes known as the *photo-electric smoke detector*, operates by means of the *light-scattering principle*. A pulsed infrared light is targeted at a photo-receiver but separated by an angled non-reflectant baffle positioned across the inner chamber. When smoke and combusted particles enter the chamber, light is scattered and reflected onto the sensitive photo-receiver and triggers the fire alarm circuit.

Some designs of optical smoke detectors draw very little current and so may be wired in conjunction with ionisation type units on the same zone. A useful operational check can be made by observing the small red light-emitting diode which flashes at intervals of 5 seconds or so.

Unlike the ionisation smoke detector, the optical type is not radioactive and is therefore completely safe when stored in vast quantities. Figure 3.17 pictorially outlines the working concept of the optical smoke detector.

Heat detector: rate-of-rise type

This type of detector responds to rapid rises of temperature by sampling the temperature difference between two heat-sensitive thermocouples or thermistors mounted in a single housing. (*A thermistor is a semiconductor whose resistance quickly decreases with an increase in temperature.*) One thermocouple monitors heat transferred by convection or radiation; the other is contained and responds only to ambient temperature. If one thermocouple increases in temperature relative to the other, an electrical potential is generated and the detector will signal the alarm sounder circuit by

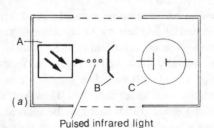

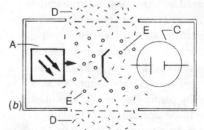

Figure 3.17. The optical smoke detector: (a) passive state; (b) alarmed state. (A, infrared light source; B, baffle; C, photo-receiver; D, smoke particles; E, scattered light.)

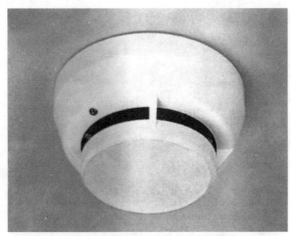

Figure 3.18. Rate-of-rise heat detector. (Reproduced by kind permission of *Photain Controls Limited*.)

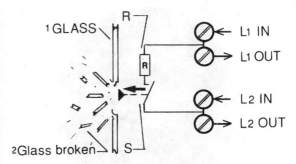

Figure 3.19. Typical fire alarm call point wiring termination arrangement. (R, resistor; S, switch.)

way of the control panel. It is wise not to install this type of unit in a boiler room or kitchen where fluctuations in ambient temperature occur regularly. This will help to avoid unwanted nuisance alarms from occurring. Figure 3.18 illustrates a typical rate-of-rise heat detector.

Heat detector: fixed temperature type

The fixed temperature heat detector is a simple device designed to activate the alarm circuit once a predetermined temperature is reached. Usually a choice of two operational temperatures is available: 60 °C and 90 °C. This type of detector is very suitable to monitor boiler rooms or kitchens where fluctuations in ambient temperature are commonplace.

Break-glass call point

Most call points may be used for either surface or flush mounting. Other than call points employed for special environments, the majority will accommodate a standard depth single-gang pressed steel socket box to BS 4662: 1970. This is an advantage when concealed work is requested.

Call points are designed to incorporate a resistor, usually valued between 470 and 680 Ω and fitted in series with the switching movement as illustrated in Figure 3.19. These are either factory fitted or connected on site, a decision made by the manufacturer. When assembled under site conditions, care must be taken to ensure that the device is correctly connected otherwise possible fault conditions could arise.

Older installations make use of small chromium-plated steel hammers to smash the call point glass. Today most units are fitted with a lightweight fragile glass served with a thin plastic membrane to protect the operative from injury sustained by broken or splintered glass. A sturdy thumb pressure is all that is required to rupture the glass and trigger the alarm sequence. This is far more convenient than using a hammer.

Testing

Regular testing may be carried out by inserting a purpose-made plastic key into a hole at the bottom of a call point. This will momentarily activate the alarm circuit. Older models are activated by releasing a small 'alan screw' housed in the top cover section of the call point.

Audio alarms

Audio alarms are known as *sounders* and can be obtained in three different forms to suit the demands of the installation. It is often the responsibility of the local fire department to advise which type should be fitted and not the electrical contractor assigned to carry out the task in hand. Should confusion arise, seek advice; it might save a lot of time and money in the long run.

In summary, sounders can be chosen from the following:

1. Fire alarm bell
2. Fire alarm siren
3. Electronic sounder

Figure 3.20. A traditional fire alarm. (Reproduced by kind permission of *Photain Controls Limited*.)

Figure 3.21. A typical fire alarm siren. (Reproduced by kind permission of *Photain Controls Limited*.)

Fire alarm bell

The traditional fire alarm (Figure 3.20) is obtainable in up to three diameters. The nominal current is between 25 and 28 mA. This provides a good audio output of between 92 and 96 decibels, measured at 1 metre. (A decibel is one-tenth of a bel and is a unit for comparing levels of sound intensities.)

Fire alarms are normally finished in stove-enamelled red and can be employed for outside use once a weatherproof gasket box has been fitted. Modern bells are polarised for fault monitoring. This means they have to be connected correctly and matched to the battery terminals. Failing to do so will make the device inoperative.

Most types of polarised alarm bells are designed using a miniaturised direct current high-speed motor. This provides operative power for a geared rotating steel striker. Older models employ the traditional 'make and break/armature' combination and are not polarised.

Fire alarm siren

The siren is often chosen for its low nominal current characteristics of between 15 and 19 mA and excellent audio output rating of between 103 and 105 decibels measured at 1 metre.

Two sizes are generally available, both of which have good working temperature ranges and will tolerate voltage fluctuations of between ±10 per cent. As with the traditional alarm bell, the siren is polarised to enable fault monitoring to be carried out. Figure 3.21 pictures a typical alarm siren.

Electronic sounder

The electric warbler is designed to cater for up to four different sounds and has a high audio output level of approximately 102 decibels. Sounders are obtainable in various colour finishes in flame-retardant uPVC but not every manufacturer provides such a colour choice.

The piezo-electric warbler is polarised to enable monitoring and is generally fitted with an internal volume control for use as and when necessary. Current consumption is between 14 and 28 mA depending on the type of unit chosen. Figure 3.22 illustrates a conventional electronic sounder.

Figure 3.22. A typical electronic sounder. (Reproduced by kind permissions of *Photain Controls Limited*.)

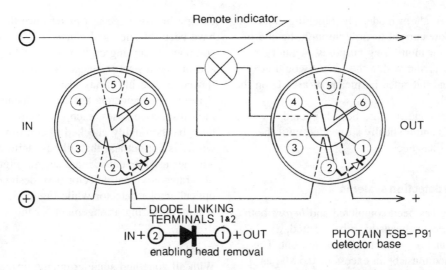

Figure 3.23. The smoke detector may be safely removed without the subsequent disconnection of other detectors and call points.

British Standards

In order to comply with the requirements of
BS 3839: Part 1: 1988, both audio alarm circuits
serving a fire detection and alarm control panel
must be used.

Where detectors are designed to be removable
from their bases, with or without locking devices,
removal of any detector(s) from the zone must not
affect the operation of any manual break-glass call
points. Reference is made to BS 5839: Part 1:
Section 6.6.2. In practice, call points are always
wired before any automatic detectors.

Leading manufacturers of fire detection
equipment, such as *Photain* and *Cerberus*, now
design fire sensor bases enabling the detector to be
safely removed without the subsequent
disconnection of other smoke alarm and call points.
Figure 3.23 illustrates how this is achieved.

Monitoring: audio alarm circuits

In order to continually monitor both cable and
terminations for open and short circuit conditions,
an end-of-line resistor, typically valued between 20
and 20 000 Ω should be connected across the
terminals of the last sounder.

As with manual call points and automatically
operated fire detectors, wiring *must* be carried
continuously with no spurs as illustrated in Figure
3.24.

Modern sounders are *polarised* meaning that the
positive conductor must be connected to the
positive terminal of the sounder and likewise with
the negative conductor. Each unit, whether it is a
bell, siren or an electronic sounder, is equipped
with a series diode. This allows effective
monitoring when a reversed polarity supply is
delivered to the sounder circuit without placing the

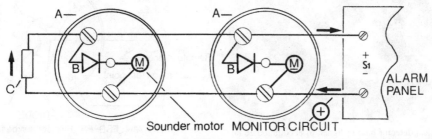

Figure 3.24. Monitoring a fire detection zone is made possible by use of an end-of-line resistor. (A, sounder base; B, diode; C, end-of-line resistor.)

sounders in an alarm mode. The blocking diode will only allow current to flow through the end-of-line resistor for monitoring purposes. A fault condition occurring within the circuit would be monitored and indicated by means of a LED on the control panel.

Once a fire condition has been triggered, monitoring is automatically suspended until the control panel has been reset.

Testing fire detection systems

When wiring has been completed and *before* both bases and detector heads have been fitted, a standard insulation test must be carried out. The value obtained must be in excess of $0.5\,\text{M}\Omega$ as recommended by the *Wiring Regulations*.

When mineral-insulated cable is used in conjunction with malleable inspection boxes, the polarised conductors can be temporarily connected to form a continuous circuit. This may then be tested at the disconnected conductors serving the control panel. If a fault condition is recorded, each individual length of cable must be inspected separately and, regrettably, any time gained will be lost.

Wiring must be carried out using either mineral insulated cable or Pirelli FP 200 grade cable. Both cables are excellent for this type of installation.

Never attempt to undertake an insulation test

using a high-voltage test meter once the base plates have been installed and connected. The electronic components serving each base together with the end-of-line resistor or capacitor could be severely flawed as a result of any test.

Before the detectors are accommodated within their respective bases, each conductor and terminal must be thoroughly checked to confirm that the wiring is correct and that all base terminal screws, whether used or unused, are secure. Figure 3.25 illustrates a typical detector base designed for an optical smoke detector while Figure 3.26 shows the schematic wiring arrangement for this base.

Final testing

With all zone and sounder circuits inspected, thoroughly tested and appliances fitted, the control panel can be connected. It is important that this is carried out strictly according to instructions provided by the manufacturer. Zones or sounder circuits left unused must have an end-of-line resistor, or sometimes a capacitor, fitted to their respective zone output terminals as shown in Figure 3.3. Most fire alarm control panels are equipped with a set of *normally open* and *normally closed* terminals. These are connected to a 'no-volt', low-voltage relay contact integrally housed with the control panel and designed to serve auxiliary equipment such as magnetic door relays. Figure 3.27 illustrates such an arrangement. An independent low-voltage supply must be provided to serve these terminals.

Confirm that *both* sounder circuits have been used in order to comply with the requirements of

Figure 3.25. Typical detector base designed to accommodate an optical smoke detector. (Reproduced by kind permission of *Photain Controls Limited*.)

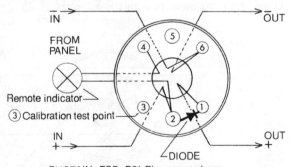

PHOTAIN FSB–P91 Fire sensor base

Figure 3.26. A schematic wiring arrangement for a typical optical smoke detector base.

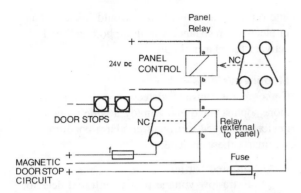

Figure 3.27. Auxiliary equipment may be controlled by means of an integral relay housed within the control panel.

BS 5839: Part 1: 1988. Should one circuit fail, the other will be there to provide an audible alarm. Finally connect the fire detectors into their respective zones making sure that any unused zones are fitted with an appropriate resistor/capacitor.

Commissioning

Once the internal batteries have been correctly connected and power supplied to the control panel, the system will be operational and the indicator lights illuminated. Commissioning is usually undertaken on a zone to zone basis. Modern manual call points can be activated by use of a specially designed plastic key which, when fitted into a small hole at the base of the accessory, engages a cam which mechanically moves the glass and, in turn, closes the terminals. This brings the system into an alarm mode. Older models are triggered by undoing the front cover using a small alan key. Closing the cover or removing the plastic test key will effectively reset the fire alarm circuit.

Testing: Ionisation type fire detector
Ionisation and optical smoke detectors can be made operative by use of a non-flammable aerosol *smoke detector tester*. This contains minute particles comparable in size to particles found in material after combustion has taken place. The aerosol tester will usually activate the smoke detector within about 10 seconds. Fortunately it is very quick to clear but it must only be used at a distance of 1 metre from the target to avoid a residue which

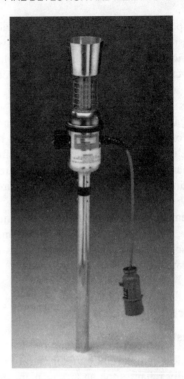

Figure 3.28. A typical heat detector tester. (Reproduced by kind permission of *Photain Controls Limited*.)

might cause damage to the device. The propellant now used is usually HCFC22 and will not contribute to environmental problems such as ozone depletion.

Heat detector test
All types and grades of heat detectors can be successfully triggered by use of a purpose-made, long-handled electric air gun. Used similarly to a hair drier, the device is targeted at the detector until the sounder circuit is activated. Most models incorporate a facility for providing cold air in order to cool the activated detector. Figure 3.28 illustrates a typical heat detector tester.

Commissioning agency
Most manufacturers of fire detection and alarm equipment provide an on-site commissioning service. The installation is tested and inspected in accordance with the commissioning demands laid down in British Standards directive 5839: Part 1: 1988. It is wise to employ the manufacturer if only

to gain peace of mind and obtain a formal commissioning certificate. A fire alarm system is a very serious and important installation which must be correctly installed and work as intended.

Records

In order to maintain a record of performance during the life of the system, a log book should be left with the customer or whoever is responsible for the maintenance and regular testing of the fire alarm installation. This complies with the requirements of BS 5839: Part 1: 1988.

Remember to leave any relevant documents, for example, guarantee cards, instruction manuals and, of course, your company trade card. This all helps to promote customer satisfaction.

Intruder alarm systems

Early systems

The earliest commercial intruder alarm installation was totally reliant upon an arrangement of electromechanical devices, which when triggered by one or more hidden switches would activate the alarm system. Cables were physically concealed for fear of damage to the circuits as monitoring and tamperproof arrangements had yet to be incorporated within the design of such an installation.

The alarm circuit was brought into play by means of simple door or window switches. Occasionally a hidden pressure switch was incorporated under a carpet at the foot of a flight of stairs. Windows and doors in commercial premises were additionally protected with internal prison type bars constructed from split steel conduit pre-wired with a delicate single extra-low-voltage cable. Cutting through the slender conduit would sever the conductor and trigger the alarm circuit, hopefully discouraging the intruders from entering!

Once electronics were marketed for commercial use, the photo-cell (*then called the electric eye*) was employed in intruder detection technology. A narrow beam of infrared light, supported by an array of concave mirrors, was arranged so as to produce a criss-cross security zone of invisible light. An intruder entering the zone, and interrupting the invisible ray, would cause a change in electrical resistance within a photo-cell and activate the alarm circuit.

Modern systems

Contemporary intruder detection systems are far more sophisticated and employ a wide range of devices designed to trigger an alarm condition. In summary, these include:

1. Infrared barrier
2. Extra-low-voltage passive infrared detector
3. Glassbreak detector
4. Pressure mat (*contacts 'normally open'*)
5. Door contacts: magnetic type
6. Personal attack panic button
7. Seismic detector
8. Motion detector: active ultrasound type
9. Body temperature detector
10. Tamper circuit (*contacts 'normally closed'*)
11. Active microwave motion detector
12. Ultrasound doppler device

A typical system would operate from an independent direct current source of between 10 and 18 volts. This is supplied from a central control panel from which all cabling serving the installation is terminated. Some intruder detection and alarm systems have incorporated a small key pad where the operator has to 'punch in' a personalised PIN in order to activate or deactivate the protected area. Other key pad facilities can be offered but choice will vary from manufacturer to manufacturer.

Figure 3.29 depicts a typical intruder alarm panel connection arrangement in which both extra-low and low-voltage mains conductors are terminated. When a low-voltage mains circuit is incorporated in an extra-low-voltage control panel a means of separation must be provided in order to segregate the two systems. Regulation 528–01–07 confirms.

Principal sounder and strobe light

It is wise to position the principal sounder on an outside wall, where it would attract maximum attention when activated.

Traditionally a bell has always played this role but today sirens and electronic horns are also employed for this important duty. Motorized sounders have a high starting current so it is

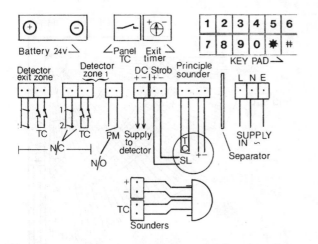

Figure 3.29. Termination arrangement serving a typical intruder alarm panel. (TC, tamper circuit; N/C, normally closed; N/O, normally open; SL, strobe light; PM, pressure mat.)

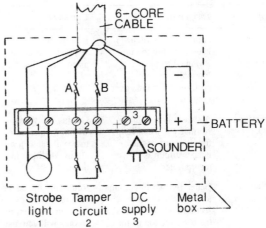

Figure 3.30. Cable termination arrangement serving a principle sounder. (A, back switch; B, lid switch.)

advisable to check and determine whether the device is compatible with the control panel relay contacts. All types of sounders have an excellent audibility factor of between 100 and 120 decibels. Internal smaller auxiliary alarms may be used throughout a protected area if necessary. These are only generally required when a building or security zone is occupied 24 hours a day. They would not, for example, be suitable for a day nursery or school but ideally suited for a premise or compound patrolled by security personnel.

A rechargeable back-up battery is installed within the principal sounder to provide emergency power if the main extra-low-voltage cable serving the device is deliberately cut.

Most burglar alarms have two *tamper switches* incorporated which will bring the system into a full alarm mode if the unit is forcibly opened or prised off the wall. Tamper circuits will be reviewed within the following paragraphs. Figure 3.30 illustrates a typical terminal arrangement serving a principal sounder. The two tamper switches shown are housed at the rear and face of the unit respectively.

Often a strobe light is incorporated forming an integral part of the principal sounder. Once triggered, this brilliant visual indicator will continue to function until the intruder alarm panel has been manually reset.

Tamper circuit

The tamper circuit, sometimes referred to as the 24 circuit, is designed to protect the wiring installation and detection devices from deliberate interference continually. All appliances are provided with a tamper circuit connected to a tiny microswitch within. Once commissioned, tamper switches are in the 'normally closed' position (Figure 3.31) and are connected in series formation with each other. Should a cover be removed from an appliance or the wiring severed a full alarm condition will automatically be triggered.

Table 3.2 briefly summarises the more practical aspects involved when tamper circuit wiring is carried out.

Passive infrared intruder detector

The passive infrared detector (PIR) operates on the principle of receiving signals from outside as, for

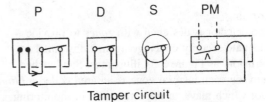

Figure 3.31. Tamper switches are wired in series formation and are 'normally closed' once the installation has been commissioned. (S, sounder; D, detector; PM, pressure mat; P, control panel.)

TABLE 3.2   Principal tamper circuit switches

Appliance	Tamper switch	Comment
Detector (passive infrared)	A tiny microswitch housed within the unit. Switched to the 'normally closed' position when the cover is fitted	Removing the cover will activate a full alarm condition
Pressure mat	A switch is not fitted. Normally a wired loop forming part of the tamper circuit is incorporated	Cutting the wires within the pressure mat will cause a full alarm condition to occur
Principal sounder	Generally two fitted. One to serve the cover section and one to serve the back section of the sounder to prevent it from being prised from its fixings	Removing the cover or forcibly removing the sounder from its fixings will activate the alarm circuit
Control panel	Fitted with a small microswitch latched to the cover section	Should the control panel be wrenched open, a full alarm state will be brought into play

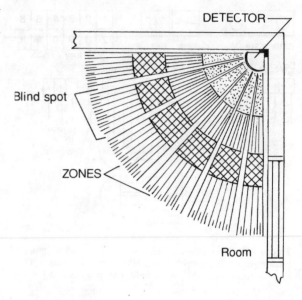

Figure 3.32. Most intruder detectors provide for a 90° to 110° cover. Between each infrared zone there is an area known as a 'blind spot'.

example, an eye would do in respect to a brain. All warm-blooded life forms radiate heat. This heat is received by the detector and compared electronically with the ambient background. The device will acknowledge this differential by converting it into a measurable electrical signal and trigger the alarm circuit.

When used as part of a security installation a PIR detector is provided with a direct current supply of between 10 and 18 volts. Ideally the device should be mounted in a suitable corner at a height of between 2 and 3 metres. Most will provide for a 90° to 110° cover, as Figure 3.32 illustrates; however, some manufacturers have claimed an even wider operating angle of 200°!

Zones

The detector is designed with *zones* to provide adequate security coverage against unwanted intrusion, as Figure 3.33 illustrates. Sandwiched between each integral zone is an area called a *blind spot* which plays an important role in monitoring detection. A radiant body moving across from one zone to another will be quickly recognised by the detector, which will electronically trigger a signal

and activate the alarm sequence. Security detectors are programmed to respond to slow-moving infrared sources such as people and to reject signals from objects like fast-moving cars.

However, many do respond to animals such as cats, dogs and cattle so it is advisable to site this type of detector *inside* a building or a secure room requiring protection.

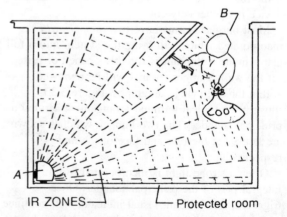

Figure 3.33. The detector is positioned to provide adequate coverage. (A, infrared detector; B, intruder.)

Polarisation

Passive infrared detectors are polarised so care must be taken to connect them correctly. A positive and negative DC supply must be wired to the relevant terminals in order to operate the device. Make a written note of the colours of the conductors and the role they play; in this way mistakes will be minimised. Current consumption is low, approximately 12 mA (0.12 A), when compared to a motorised sounder so standard extra-low-voltage multi-core cable can be used for installation purposes.

Radio interference

Not all PIRs are protected or screened from radio interference and this should be borne in mind at the design stage of the installation when options for choice are made. Installing an intruder detection and alarm system too near a radio transmitter could have disastrous consequences as certain frequencies can trigger an alarm condition.

Ranges of detection

Many PIRs have a built-in option of between one and three ranges of detection. This is made possible by either changing the plastic lens positioned in front of the detector or by physically moving the integral printed circuit board up or down to align with the appropriate fixed lens. These options will,

of course, vary from manufacturer to manufacturer, but if in doubt, read the instruction leaflet!

Figure 3.34 shows a cross-sectional side view of infrared coverage expected from an average PIR detector using a long-range lens.

Connections

A typical PIR detector will accommodate up to three circuits. A small connector block provides two terminals for a permanent DC extra-low-voltage supply and a futher two for the alarm circuit. The last pair of terminals are reserved for the series-wired tamper circuit. It is essential to connect each conductor in the terminal block correctly since damage could result from wiring incorrectly. Figure 3.35 illustrates a typical terminal connection arrangement serving a passive infrared detector.

Wiring techniques for intruder alarms

There seems to be an underlying but understandable reluctance among less-experienced electricians to undertake any form of intruder alarm installation. It is no more difficult than any other comparable type of electrical work; only different.

The *Wiring Regulations* do not apply to intruder alarm systems as the source of power serving the installation is by means of an isolating transformer,

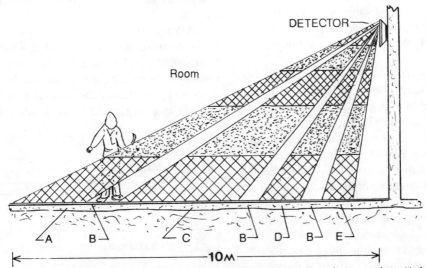

Figure 3.34. A side view of the coverage expected from an average PIR detector using a long-range lens. (A, long-range area 10 m plus; B, blind spot; C, transitional area 5–10 m; D, short-range area 3–5 m; E, prowl area 1–3 m.)

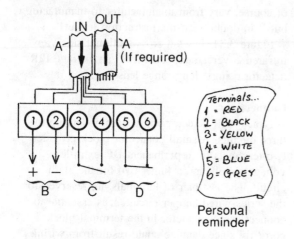

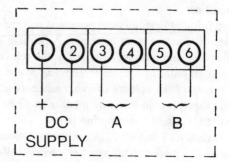

Figure 3.36. A typical PIR intruder detector cable termination arrangement. (A, alarm circuit; B, tamper circuit.)

Figure 3.35. Cable termination arrangement serving a typical passive infrared detector. (A, flexible six-core extra-low-voltage cable; B, supply voltage to detector; C, alarm circuit; D, tamper circuit.)

where the secondary winding is completely isolated from earth. However, requirements are laid down for the physical segregation of such circuits from other services such as low-voltage mains, fire alarm and emergency lighting installations.

The *Institution of Electrical Engineers* classify an intruder alarm system as a *Category 2 installation*. This is further reinforced by Regulation 528–01–03 and BS 6701, which demand that extra-low-voltage cabling must be segregated from other electrical services. When multi-core cable is used adequate insulation must be afforded and supplementary installation added where hazardous conditions occur.

The *Wiring Regulations* permit Category 2 cables to be laid in the same ducting or drawn into conduit as those occupied by Category 1 circuits. However, the level of insulation serving the Category 2 cables must be at least equal to that of the highest voltage present (Regulation 528–01–01). In practice, it is wiser to install multi-core extra-low-voltage flexible cable laid in mini-trunking or drawn through an independent uPVC conduit. If, for practical reasons, this cannot be done, distance all cabling from low-voltage mains and Category 3 circuits. Avoid cable runs close to large-current-consuming electric motors or contactors and also avoid the use of telephone wire.

Zoning

Extra-low-voltage circuits serving intruder detector devices are known as *zones*. In a simple arrangement each zone is served with a number of passive infrared detectors, or *contacts normally closed* devices. All PIR are placed in parallel formation in one continuous circuit. Spurring or branch circuits are not permitted as monitoring would be possible.

Usually a six-core flexible cable is required to provide a direct current supply, alarm and tamper circuit to each detector. The tamper circuit is connected in *series* formation as shown in Figure 3.31, whereas the supply and alarm circuits are wired independent of each other in *parallel* formation. Figure 3.36 illustrates a typical PIR terminal arrangement (*see also 'Connections' above*).

Cable runs should be kept as short as possible. Many manufacturers recommend that each zonal loop should be no greater than 10 ohms in resistance. In practice this value may be quickly assessed by measuring the resistance of 1 metre of looped cable (Figure 3.37) and dividing the product

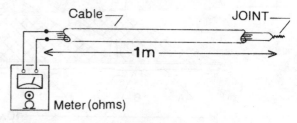

Figure 3.37. Measure the resistance of 1 m of looped intruder alarm cable.

into 10 ohms. The resolution will provide for the maximum length permitted. This is best carried out using a digital milli-ohmmeter.

The timed zone

The intruder detector zone protecting the control panel and designated exit route is known as the *timed zone*. All wiring associated with this area must be terminated in the dedicated terminals provided within the control panel (see Figure 3.29). Upon arming the system, an internal buzzer is activated advising that the operative has, typically, between 10 seconds and 2 minutes to evacuate the premises by means of the designated route. The timed zone will prevent the alarm sequence from triggering once the panel has been mobilised. The timing should be adjusted to meet with the requirements of the operative charged with arming or disarming the system on a daily basis.

Principal sounder

The principal sounder is usually placed externally and in a position which, when activated, would draw maximum attention. Wiring is usually carried out using either 6- or 8-core extra-low-voltage flexible cable. The design of the system will dictate the number of conductors required.

As with the PIR detector, the principal sounder is served with more than one circuit. This has been considered in a previous paragraph. (*See also Figure 3.30.*)

Electrical mains supply

As an intruder alarm system consumes very little current from the electrical mains, a lighting grade cable would be of sufficient size to serve the control panel. This must be a dedicated circuit originating from the main switch-gear and terminated adjacent to the control panel with an unswitched fused connection unit fitted with a 3 or 5 amp fuse. Figure 3.38 illustrates this concept more clearly.

Testing the alarm system

Extra-low-voltage cables should be tested using a simple continuity test meter before any detection devices are terminated. On no account must testing

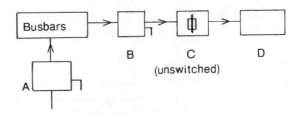

Figure 3.38. A dedicated supply serving an intruder alarm panel. (A, main switch; B, dedicated MCB; C, switch-fused connection unit; D, control panel.)

be carried out using a high-voltage insulation tester. After all detection appliances and accessories have been fitted, terminated and checked, verify that all lids and cover plates served with tamper circuits are firmly in place. Covers that are not properly fitted will bring the alarm circuit into play.

Once visual and instrumental testing has been completed and all devices have been fitted it is wise to warn site operatives that a commissioning programme, as recommended by the manufacturer, is imminent.

Remember to leave relevant documents, for example, log book, instruction manual and guarantee card with your customer.

Emergency lighting installations

The *Wiring Regulations* classify emergency lighting arrangements as *Category 3 circuits* which must be supplied from one of the following safety sources:

- Electrochemical (batteries)
- Diesel-driven generator

The governing regulations

Extra-low-voltage lighting systems are regulated by recommendations laid down in the *Wiring Regulations* (Regulation 110–01–01) and *British Standards directive 5266*; a copy of which may be borrowed from most leading public libraries.

Cables serving self-contained emergency lighting, for example, maintained, non-maintained or sustained luminaires are not considered as emergency lighting circuits. An authentic circuit will originate from a central battery system or diesel generator unit and will be wired using mineral insulated cable or a cable such as *Pirelli*

FP 200 which complies with the requirements of BS 6387.

Installation requirements

Emergency lighting circuits *must never* be laid into the same ducting or drawn into the same conduit as those occupied by Category 1 circuits. *British Standards directives 5266 and 5839* together with the *Wiring Regulation* 528–01–04 recommend that cables serving an emergency lighting installation are completely segregated from cables of any other category. When installed in trunking with other services, Category 3 cables must be discriminated from other conductors by a continuous partition as demanded by Regulation 528–01–06. However, this directive also confirms that if mineral insulated cable or cable complying with the requirements of BS 6387 (*Performance Requirements for Cables required to maintain Circuit Integrity under Fire Conditions*) are used, then partitioning is not normally required.

In no circumstances must emergency lighting circuits be contained within a common multi-core or flexible multi-core cable which also contains Category 1 low-voltage mains circuits.

Keep Category 3 circuits isolated from other services.

Classification of emergency lighting

Emergency lighting can be divided into five main groupings:

1. Maintained
2. Non-maintained
3. Sustained
4. Central battery
5. Diesel generator

Maintained

This type of luminaire is supplied with a single light source which may be switched on and off as needed. Should the mains or local over-current protection fail, the emergency light will automatically switch on, powered from its own rechargeable batteries. Switching on or off only interrupts the light source and not the mains supply to the battery unit.

Non-maintained

Should either the subcircuit or main electricity supply fail, the luminaire will automatically switch on, powered by its integrally fitted rechargeable batteries.

Sustained

Two independent light sources are fitted within a sustained emergency lighting unit. One source relies upon main electricity and may be switched on and off as required. The other will only operate through mains or local circuit failure and power is supplied remotely or by integrally fitted rechargeable batteries linked to an electronic inverter. Regulation 561–01–01 caters for this type of emergency light. Sustained, maintained and non-maintained lighting arrangements must always be connected to a neighbouring sublighting circuit in order to protect against total darkness.

Central battery

This type of system typically comprises a specified number of lead acid or nickel cadmium secondary cells housed in a cabinet and connected in a series formation to provide the desired working voltage of either 25, 50 or 110 volts. Mineral-insulated cables or cables complying with the recommendations of BS 6387 are used for circuits throughout the protected area.

Volt drop

Volt drop must be seriously considered, especially when dealing with long runs of mineral-insulated cable. Expression [3.1] may be used to calculate the expected drop in voltage due to the intrinsic resistance offer by the conductor.

$$\text{Volt drop} = \frac{(\text{Tabulated mV}/I/\text{m}) \times l}{1000} \qquad [3.1]$$

where $\text{mV}/I/\text{m}$ is the tabulated millivolt per amp per metre in the conductor considered (Regulation Table No. 4J1B), and
l is the length of the cable run in metres.

As an example, consider the following:

Calculate the volt drop which will occur in 40 metres of 1.5 mm² twin mineral-insulated cable supplying 10 in number, 15 watt incandescent emergency lights from a 110 volt central battery system. For the purpose of this example correction factors may be ignored.

First the design current I_b must be found.

$$\text{Design current } I_b = \frac{\text{Total power in watts}}{\text{Applied voltage}} \qquad [3.2]$$

Substituting for known values:

$$I_b = \frac{10 \times 15}{110}$$

$$= 1.36 \text{ amps}$$

Referring to the *Wiring Regulations* (mineral-insulated cables): a twin conductor with a nominal cross-sectional area of 1.5 mm² produces a volt drop of 30 mV per amp per metre. Applying Expression [3.1] and substituting tabulated figures,

$$\text{Volt drop} = \frac{30 \times 1.36 \times 40}{1000}$$

$$= \frac{1632}{1000}$$

$$= 1.632 \text{ volts}$$

Regulation 525–01–02 stipulates that voltage drop between the origin of the installation and the nearest point of utilisation should not exceed 4 per cent of the nominal voltage. This may be quickly calculated by use of the following expression:

$$\text{Permissible volt drop}$$
$$= \text{Nominal voltage} \times 0.04 \qquad [3.3]$$

Substituting known values,

$$\text{Permissible volt drop} = 110 \times 0.04$$
$$= 4.4 \text{ volts}$$

The volt drop resolved in this example is satisfactory and complies with general requirements. Being only 1.632 volts, it represents 1.79 per cent of the terminal voltage.

Over-current protection: batteries
Central battery systems must be protected against

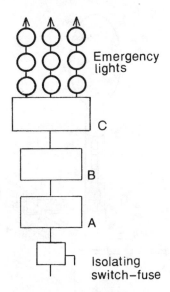

Figure 3.39. Central battery emergency lighting wiring arrangements. (A, charger/control; B, batteries; C, distribution centre.)

over-current by means of circuit breakers or cartridge fuses to BS 1361. It is wise to be generous when consideration is given concerning the number of circuits that will be used as the effect of volt drop throughout long cable runs must never be ignored. Figure 3.39 illustrates a typical arrangement serving such a system.

Batteries must be rated so as to provide a minimum of 3 hours of power, be compatible with the charger unit and capable of recharging to 90 per cent within 12 hours. It is advisable to physically inspect the batteries regularly and test the terminal voltage of each cell. A check should also be made to see that each unit is properly topped up with distilled water and that the relative density of the electrolyte (*the acid and distilled water combination*) is as recommended. Figure 3.40 illustrates a simple hydrometer, an instrument for determining the relative density of a liquid.

Relative density can be defined as the ratio of the density of a substance to that of water.

Load test
A 3 hour load test should be implemented enabling lamps to be checked and illuminosity values measured at the commencement and close of the test period. This may be carried out by use of a

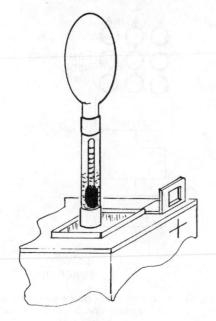

Figure 3.40. A simple hydrometer for measuring the relative density of the battery acid.

Figure 3.41. A typical log book page to record formal emergency lighting tests.

Page No. 8

TEST RECORD
emergency lighting K.F LIMITED

LUMINAIRE LOCATION. 'B' STAIRS (TOP)
REF·No. EM8

Type of test	1st year Signed & date	2nd year Signed & date	3rd year Signed & date	4th S
FUNCTIONAL	CL 1.6.99			
FUNCTIONAL	PL.t 2.6.99			
ONE HOUR DURATION	SL1. 1.12.99			
FUNCTIONAL		Elij. .3.1.01		
FULL DURATION		Li. 1.6.01		

suitable light meter. Values resulting from formal tests should be recorded in a log book and kept for inspection. Figure 3.41 illustrates a typical page.

Generators

Generators are often used as a source of energy to supply power to selected luminaires for the purpose of illuminating open areas and escape routes upon mains failure. Providing emergency power is totally dependent upon the starter supply, so batteries chosen must be of the same industrial standard as that demanded for central battery emergency lighting systems.

Generators must be locally protected by use of fuses or suitably designed and rated miniature circuit breakers. Generally, automatic changeover from mains electricity to locally generated power should occur within one second of mains failure.

This system is ideal for standby lighting arrangements but consideration must be given to the kilovolt/ampere (kVA) rating (*maximum output current*) of the generator. It is easy to inadvertently add more luminaires. An increase in current demand would lead to an overloaded generator and could cause problems to both generator and installation alike.

Positioning emergency lighting

The European Community directive concerning emergency lighting requires a minimum of 1 lux for specified points of emphasis and areas of normal risk, whereas unobstructed escape routes need only be 0.2 lux to conply with this directive.

The term *lux* (symbol 'lx') is a *Système International* (SI) unit of illuminance representing one lumen per square metre.

Any building where people are employed must comply with the emergency lighting and exit sign requirements. Escape routes and open areas such as lobbies, classrooms, offices, etc., must be provided for and suitable exit signs installed.

Figure 3.42 depicts a selection of illuminated signs now incorporated in new building projects.

For guidance only, Figure 3.43 illustrates a typical emergency lighting arrangement. Luminaires should be prudently located so that fire extinguishers, alarm call points and fire-retardant doors are served with adequate levels of light. It is wise to seek guidance from the appropriate fire authority concerning the final positioning of the emergency lighting units. This reduces embarrassment and the risk of remedial work

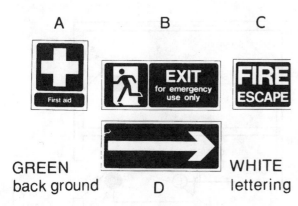

Figure 3.42. Safety signs complying with BS 5378, Part 1. (A, first aid post; B, emergency exit; C, fire escape; D, escape route.)

having to be carried out after the installation has been inspected!

High-risk areas

In high-risk areas *European Standards* require that at least 10 per cent of the normal illumination is available within 0.25 second from electrical mains failure. Typically, these areas may be defined as:

- acid baths
- conveyors
- dangerous areas
- rotating machinery
- turnstiles

The level of lighting required can be achieved successfully by use of non-maintained or sustained fluorescent luminaires. Sustained fittings incorporate an *inverter* which produces 230/240 volts alternating current upon mains failure. Battery packs serving an inverter are integrally fitted within the unit or sometimes sited remotely. They are designed to provide up to 3 hours of emergency power.

Specific locations

A lighting level producing a minimum of 1 lux of illumination sustained for a minimum of 3 hours is required in specific locations. To explain further, these locations may be defined as:

- bars or areas selling alcoholic drinks
- escalators
- lifts
- plant rooms
- staircases
- toilets
- workshops

General locations

European Standards require that maintained emergency lighting is provided in premises where the general public could be unfamiliar with the design or internal arrangements of a building. In

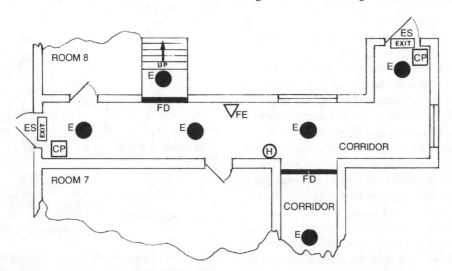

Figure 3.43. Each section of an escape route must be provided with at least two emergency lighting points to guard against total darkness. (CP, call point; E, emergency light; FD, fire door; FE, fire extinguisher; ES, exit sign pictogram; H, first aid point.)

practical terms this refers to hotel and airport lounges, bars, exhibition halls, stately homes, etc. Other areas include:

- Cinemas and theatres
- Classroom and community areas
- Emergency exits
- Escape routes
- Areas adjacent to fire doors
- Offices and public buildings
- Open areas such as restaurants, libraries, etc.
- Shops and superstores
- Where a change of direction is experienced in an escape route

Provision of supply

Maintained, non-maintained and sustained emergency lighting luminaires must be supplied from a lighting subcircuit serving the area where the emergency luminaire is located.

Figure 3.44 illustrates this concept more clearly. In no circumstances should emergency lighting be supplied from a power circuit or a lighting circuit serving a remote area. Upon mains or submains failure, emergency lighting affiliated to the area affected must be capable of automatically switching on in order to protect against total darkness.

Central battery system

The central control panel should be provided with a dedicated supply originating from the main electrical distribution centre (Figure 3.45). The disadvantage of this method is that emergency facilities can only be activated upon total mains failure. However, it is possible to use the central battery system during a fire alarm condition by utilising a set of *auxiliary 'no-volt' contacts* within the fire alarm panel. These 'normally closed' contacts provide a means to control and energise an externally located relay, the switching contacts of which are wired so as to maintain power to the emergency lighting panel. A signalled fire condition will automatically open the 'normally closed' auxiliary contacts and drop the external relay. Once de-energised, the external relay contacts open, isolating the mains supply to the emergency

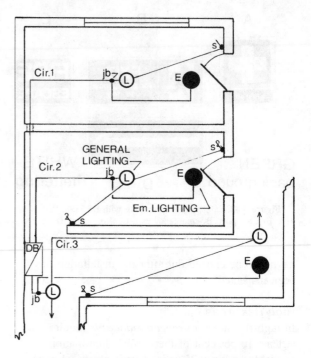

Figure 3.44. Each area is served with a non-maintained emergency light supplied from the lighting circuit serving the room. (JB, joint box; DB, distribution centre; L, mains lighting point; E, emergency light; S, switch.)

lighting control panel, The automated panel then provides a direct current supply to the emergency lighting circuits. Figure 3.46 outlines, in schematic form, details of this arrangement.

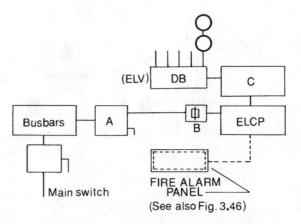

Figure 3.45. Typical arrangement to supply mains voltage to a central battery emergency lighting system. (A, control switch-fuse; B, local switch-fused connection unit; C, batteries; ELCP, emergency lighting control panel; DB, distribution board.)

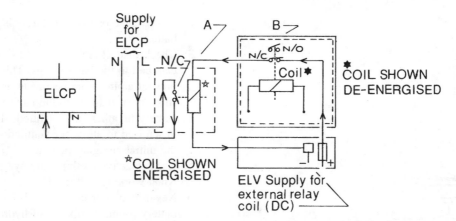

Figure 3.46. An emergency lighting arrangement can be automatically switched on whenever a fire alarm condition is registered. (ELCP, emergency lighting control panel; A, external relay; B, fire alarm control panel; N/O, normally open; N/C, normally closed.)

Testing emergency lighting

Self-contained luminaires

As a practical guide, emergency lighting units equipped with rechargeable batteries should be charged as follows:

- Initial commissioning charge of 60 hours.
- Recharging after a 3 hour test period to a minimum of 24 hours.

Once the initial charge has been carried out it is useful to write the commissioning date on the battery label. This will help others who follow when they are making technical decisions.

BS 5266 recommends that the red light-emitting diode serving the luminaire should be checked on a daily basis. This information should therefore be offered to the client.

Self-contained emergency lighting units must be tested regularly, and as a guide the following recommendations should be passed on to the customer:

- Monthly: switch on for 5 minutes.
- Six monthly: switch on for 1 hour.
- Three yearly: full duration of 3 hours.

Central battery systems

A full insulation test must be carried out; the resultant value of which must not fall below 0.5 megohms.

With all lamps removed and the distribution centre completely isolated from the direct current supply, an insulation test should be made between both conductors and between conductors made common and the copper cable sheath. If Pirelli FP 200 cable is used, then the test can be made between the conductors and the integral tinned copper circuit protective conductor. Figure 3.47 illustrates how this test may be implemented in practical terms when using mineral-insulated cable.

Once completed, a 3 hour load test should be put into effect and a check made with a digital lightmeter to confirm that levels of illumination are as required and the system is working as designed.

Health and safety

Sustained emergency luminaires contain an electronic device known as an *inverter*. This will automatically produce a 230/240 volt supply upon submains or mains failure. Although currents from the device are not normally lethal, inverter circuits can produce an unpleasant or painful electric shock. This might prove dangerous to some extent if, for example, service work is being undertaken on a pair of steps or high ladder. It is wise to disconnect the batteries before starting repairs or maintenance work.

Nickel cadmium: rechargeable batteries
Provided sensible precautions are carried out, the risk of battery contamination is greatly reduced. Listed at random is a selection of the more

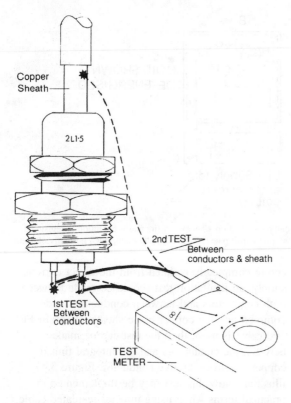

Copper Sheath

2L1·5

2nd TEST
Between conductors & sheath

1st TEST
Between conductors

TEST METER

Figure 3.47. An insulation test must be carried out of which the resultant value should not fall below 0.5 megohm.

important safety aspects which should be remembered when handling batteries:

1. Batteries past their useful life must be removed from the luminaire.
2. Avoid skin contamination from leaking batteries.
3. Never dismantle or puncture.
4. In no circumstances give exhausted batteries to children. Dispose of batteries sensibly.
5. Never incinerate exhausted batteries. They could explode.

Secondary cells: serving a central battery unit
The following guidelines are to provide a practical insight into the many potential dangers and hazards that are often present when emergency lighting battery units are assembled for use:

1. Handle with care when assembling and observe the correct potential.

2. Never leave or place metal tools on top of a battery.
3. Check that all connecting links are securely clamped to the battery lugs. This will prevent arcing from occurring and subsequent damage to the battery.
4. Battery breather caps, if fitted, should be unscrewed to allow for ventilation during the initial charging period.
5. Use protective clothing and eye protection when working with acid.
6. Never smoke or use a naked flame in a battery room. A mixture of hydrogen and oxygen is given off during the charging process and is highly inflammable.
7. Make sure that the battery room is well ventilated.
8. Never short circuit the terminals of a battery. The voltage may be low but the current drawn can be very high. Arcing can damage the battery plates and cause an explosion.
9. Rings and watches should be removed from the hands and wrist before using metallic tools on secondary cells. Accidental connection with personal jewellery could cause skin burn to occur.
10. Acid spillage should be flushed away with plenty of water.
11. Should acid spillage contaminate the skin, dilute with plenty of clean water. Seek medical advice.
12. If acid is accidentally swallowed, advise victim to drink plenty of clean, fresh water. Seek immediate medical attention.
13. Always wash your hands thoroughly before eating or drinking.

Keep to the guidelines and remain safe from danger!

Nurse call system

In its simplest form a *nurse call system* is designed to register a call for assistance from a sick room. Summoning help is made by simply pressing a button or pulling a retractive cord switch.

Once the call has been registered an over-door indicator lamp provides means of location from the

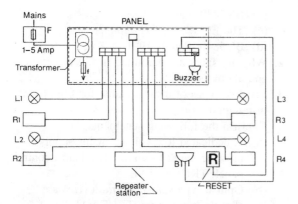

Figure 3.48. Simple nurse call system. Wiring is carried out using extra-low-voltage flexible cable. (R1 to R4, room call units; L1 to L4, 'over-door' indicator lamps; B, remote warning buzzer; F, fused connection unit.)

respective corridor. Alternatively, a local paging system can be brought into play. In either case once the demand has been addressed the alarm may be manually cancelled at the control centre or local repeater panel.

Nurse call systems may also be adapted to other applications such as sheltered accommodation for the elderly, the monitoring of external fire doors, means of providing a personal alarm in case of attack or may be used to serve police station detention cells.

Figure 3.48 illustrates a basic mains-operated, battery standby nurse call system in block diagrammatic form which would be ideally suited for a small department in a private hospital.

Summary

Fire detection and alarm systems

1. Regulations and requirements governing the installation of fire detection and alarm systems must comply with the demands of BS 5839: Part 1: 1988.
2. Detectors and sounders must be wired separately in a continuous parallel formation arrangement. No spurs.
3. The installation must be segregated from other installations.
4. The system is wired using either mineral-insulated cable or a suitable fire-resistant and flame-retardant low smoke halogen cable.

5. A dedicated low-voltage mains circuit must be wired to serve the fire alarm control panel.
6. Devices which may be incorporated within the design of a fire detection and alarm system are as follows:
   (a) Smoke heads (ionisation and optical types)
   (b) Heat detectors (rate-of-rise and fixed temperature type)
   (c) Call points (manual)
   (d) Sounders (bell or siren)
   (e) Beam detectors
   (f) Duct detectors
7. The smoke detector, now formally known as the *automatic ionisation fire detector*, was developed in Bad Ragaz, Switzerland, by the Cerberus Company in the early 1950s.
8. Both call points and automatic detectors are constantly monitored by use of an end-of-line resistor or capacitor placed in parallel with the last device.
9. Testing a smoke head should be carried out using a non-flammable aerosol smoke detector tester.

Intruder alarm systems

10. Modern intruder detection installations employ a wide range of devices. Examples are:
    (a) Active and passive infrared detectors
    (b) Infrared detectors
    (c) Seismic detectors
    (d) Glass-break detectors
    (e) Ultrasound and microwave motion detectors
    (f) Body temperature detectors
11. A typical system operates from an independent direct current source of between 10 and 18 volts.
12. An installation is wired using extra-low-voltage multi-core flexible cable.
13. The tamper circuit is wired to protect the installation from deliberate interference.
14. Not all passive infrared intruder detectors are protected or screened from radio interference.
15. Circuits serving detectors originating from a central control panel are known as zones. The intruder detector zone protecting the control panel is known as the timed zone.

16. A dedicated circuit originating from the low-voltage mains must be provided to serve the control panel.

Emergency lighting installations

17. Emergency lighting arrangements are classified as *Category 3 circuits*.
18. Emergency lighting circuits should never be mixed with *Category 1 circuits*.
19. Emergency lighting systems are classified as follows:
    (a) Maintained
    (b) Non-maintained
    (c) Sustained
    (d) Central battery
    (e) Diesel generator
20. Volt drop must be considered, especially when long runs of mineral-insulated cable are used.
21. In high-risk areas *European Standards* require that at least 10 per cent of the normal illumination is available within 0.25 second of mains or circuit failure.
22. Maintained, non-maintained and sustained emergency luminaires must be supplied from a lighting subcircuit serving the area where the emergency light is located.
23. A central battery control panel must be provided with a dedicated supply originating from the low-voltage mains.
24. Sustained emergency lighting luminaires are equipped with an inverter which produces 230/240 volts supply upon mains or circuit failure. Disconnect the battery before servicing the fitting.
25. It is wise not to place tools on top of a secondary cell or smoke in a battery room. A mixture of hydrogen and oxygen is given off during the charging process and is highly inflammable.

Review questions

1. Regulations and requirements governing the installation of fire detection and alarm systems must comply with the demands of one of the following:
   (a) The *IEE On Site Guide*
   (b) BS 5839: Part 1: 1988
   (c) The EC *Wiring Regulations*
   (d) BS 5840: Part 1: 1991
2. Why are spurs not permitted in fire detection and alarm circuits?
3. List two types of cable suitable for fire alarm installation work.
4. Confirm the following statements:
   (a) A dedicated low-voltage mains supply must be wired to serve a fire alarm control panel. TRUE/FALSE
   (b) Optical type smoke alarm devices are slightly radioactive. TRUE/FALSE
   (c) The automatic ionisation fire detector was developed by the *Cerberus Company*. TRUE/FALSE
   (d) Electronic sounders may be wired using any type of cable. TRUE/FALSE
5. Why are modern sounders polarised?
6. Briefly describe the role of the tamper circuit in an intruder detection and alarm system.
7. Confirm the following statements:
   (a) Passive infrared detectors used in intruder alarm circuits will detect movement through glass. TRUE/FALSE
   (b) Intruder alarm installations are classified as Category 3 systems. TRUE/FALSE
   (c) Installation cables should be kept as short as possible. TRUE/FALSE
   (d) Extra-low-voltage cables should be tested with a simple continuity tester. TRUE/FALSE
8. Why is it unwise to install an 'off the shelf' intruder alarm system too near a radio transmitter?
9. Describe the protection afforded against unwanted tampering with an intruder alarm panel.
10. What is the name given to the intruder detector zone protecting the control panel?
11. What classification is given to emergency lighting circuits in the *Wiring Regulations*?
12. Name a suitable type of cable in which to wire an emergency lighting system originating from a central battery unit.
13. List three areas which are considered to be of 'high risk' where at least 10 per cent of the level of normal illumination must be available

within 0.25 of a second from mains or circuit failure.

14. State two practical methods of overcoming volt drop when mineral-insulated cable is used to serve an emergency lighting system from a central battery unit.

15. Each section of an escape route must be provided with at least N number of emergency light to guard against total darkness. N is equal to one of the following:
    (a) Three
    (b) One
    (c) Four
    (d) Two

Handy hints

1. Strong radio transmissions stemming from commercial VHF wavelengths can activate passive infrared detectors serving intruder alarm installations and security lighting arrangements.

2. Beam-type fire detectors should not be installed where powerful radio transmissions are likely.

3. Never dismantle an ionisation-type smoke detector as it contains a small amount of the radioactive element *americium 241* (approximately 0.9 microcuries).

4. PIR units designed to be incorporated within an intruder alarm installation are factory fitted with a tiny microswitch. Care should be taken when removing or reinstating the cover of the passive infrared detector to avoid damage to the microswitch.

5. Adopt a common cable-marking policy when two or more electricians are working together. This will avoid confusion at the second fix stage of the installation.

6. Maintain a good standard of workmanship when installing intruder alarm systems. Never route cables under carpets nor clip them to the top side of floor boards. Keep all Category 2 circuits well away from low-voltage mains cabling.

7. Class 3 appliances and equipment should never be fitted with plugs suitable for use with low-voltage mains installations. Always use a non-standard plug so that confusion will not arise.

4 Fundamental requirements

In this chapter: The kilowatt-hour meter, the simmerstat, types of thermostats, thermometers, thermoelectric thermometer, gas thermometer. Linear expansion, principle of moments, and centre of gravity. Conditions of equilibrium, principles of levers, defining work done, mass weight and force. Efficiency of a machine and wasted energy.

Fundamental requirements is an assortment of topics which, if related singly, would fail to provide sufficient facts and information to compile a full chapter. Each issue is a requirement of the revised *City and Guilds 236 Electrical Installation course* or a topic relating to *National Vocational Qualifications* in Electrical Installation.

The kilowatt-hour meter

Dial type

When electricity was first commercially consumed, costing was rather casual and based on the number of power outlets and lighting points installed. This system eventually proved commercially impractical and provided an opening for the introduction of prepayment and credit meters.

For explanatory purposes the kilowatt-hour induction meter can be considered as a skeletal electric motor in which the armature is formed from a non-ferric disc (Figure 4.1(a)). Current consumed passes through the windings of an electromagnet incorporated within the meter which

(a)

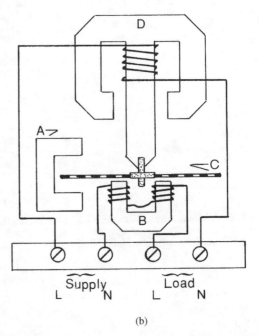

(b)

Figure 4.1. (a) The kilowatt-hour meter; (b) principle of the kilowatt-hour meter. (A, magnetic brake; B, current coil; C, rotating disc; D, voltage coil.)

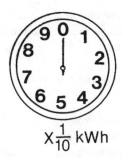

$$X\frac{1}{10} \text{ kWh}$$

Figure 4.2. A complete revolution of the smallest dial represents one unit of electricity consumed.

provides power to rotate the disc (Figure 4.1(b)). In order to obtain a record of the amount of current consumed, the revolving disc is geared to an arrangement of interconnecting cog wheels designed so as to enable the meter to record kilowatt-hours. The speed of the disc is directly proportional to the quantity of current passing through the meter and is further controlled by means of a large permanent magnet circumnavigating a small section of the disc.

Dial movement

One revolution of the first dial graduated in units will advance the second dial, which measures tens, by one division. Once the second dial completes a full cycle it mechanically moves the third dial, which measures hundreds, by a single division and so on.

A completed revolution of the smallest dial represents one *unit* of electricity used (Figure 4.2). The consumption of electricty is measured in *kilowatt-hours* or *units* for the purpose of costing and equals *1000 watts of power used in a period of one hour*. When working in *Système International d'Unités* (SI units) it represents *3.6 megajoules*. A *joule* can be defined as *the work done per second by a current of 1 amp flowing through an impedance to the value of 1 ohm.*

Reading a kilowatt-hour meter

Dial meter

Figure 4.3 represents a typical dial face arrangement serving a single-phase kilowatt-hour meter. The dials are divided into units of 10 000, 1000,

100, 10 and 1 to record the amount of electricity consumed. The smallest dial, often coloured red with white figures, is employed for testing purposes and can be ignored when reading the meter.

First target the dial recording the highest value, which in Figure 4.3 is ranged bottom left and calibrated in divisions of 10 000 units. Take note where the dial indicator is aligned and select the lower of the two figures either side of the dial indicator. Figure 4.3 shows that the indicator has passed 1. Now select the next highest valued dial, located above the bottom left dial and take the smallest value of the two figures ranged either side of the dial indicator. In this case it is 3. Review the last three dials by repeating the procedure ensuring that values are taken in strict dial-value sequence. The kilowatt-hour meter will then read 13 485 units.

Digital meter

With this type of kilowatt-hour meter the amount of electricity consumed is displayed as a single row of numbers which is read directly from the meter. When restricted (off-peak) consumption is recorded within the same meter, two independent rows of figures are used; one reserved for day tariff units consumed, the other for the lower priced economy rate.

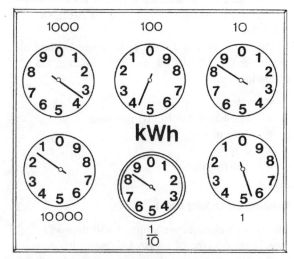

Figure 4.3. A typical dial face serving a single-phase kilowatt-hour meter.

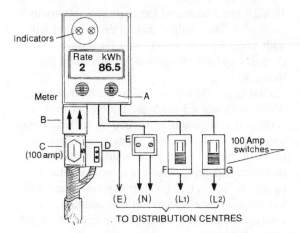

Figure 4.4. Graphical arrangement of a telemeter serving a restricted and unrestricted supply. (A, panel display control buttons; B, security cover; C, service fuse; D, neutral and PME termination box; E, neutral termination box; F, unrestricted supply miniature isolator; G, restricted supply miniature isolator.)

Radio telemeter

Figure 4.4 graphically illustrates a modern radio telemeter fitted by local supply authorities to provide metering for both restricted and unrestricted single-phase services to typical homesteads.

The meter has incorporated a miniature FM radio receiver/switch which responds to signals from BBC Radio transmitters at preset times of the day. This effectively switches the restricted section of the telemeter to provide cheap rate electricity in the home. The 24 hour unrestricted tariff is constantly in circuit, recording whenever current is drawn from the supply.

The liquid crystal display panel ranged centre of the meter provides, at the touch of a button, the following data:

- rate
- kilowatt-hours (units) recorded
- time of day
- date
- tariff or status

Costing electricity used

The cost of electricity consumed will depend largely on the type of contract arranged between the supply company and the consumer and will vary depending on whether an industrial,

commercial, agricultural or domestic tariff has been agreed. As a practical example, consider the following problem:

Calculate the cost of the following domestic account:

Previous meter reading:	11 908
Present meter reading:	13 488
Quarterly standing charge:	£11.35
Cost per unit:	9.2 pence
Value Added Tax:	8 per cent

First the total number of units consumed must be found.

$$\text{Units consumed} = \text{Present reading} - \text{Previous reading}$$

$$= 13\,488 - 11\,908$$

$$= 1580 \qquad [4.1]$$

$$\text{Cost of units at 9.2p per unit} = \frac{9.2 \times 1580}{100}$$

Cost of units	= £145.36
Standing charge	= £ 11.35
Total cost	= £156.71
Value Added Tax	= £ 12.53
	(that is, 8 per cent of £156.71)
Grand total	= £169.24

Simmerstat

Application

A simmerstat is used to regulate heating loads. Every home which has a modern electric cooker is equipped with at least four or five simmerstats serving to provide temperature control to individual radiant plates and a single grill element (Figure 4.5).

A domestic simmerstat's industrial counterpart is often known as a *power controller* or *burst firing module* and is capable of switching far greater loads. Basically these operate in a similar fashion to their domestic cousins although some apply solid-state electronics in order to function. Generally they are employed for temperature control and often serve package-sealing machinery found in industry.

Figure 4.5. A domestic simmerstat.

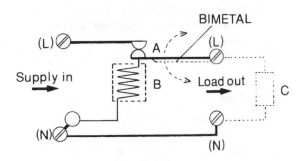

Figure 4.6. The switching and wiring arrangements of a domestic simmerstat. (A, control contacts; B, internal heating element; C, external heating load.)

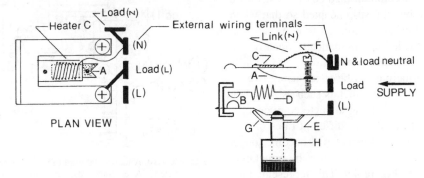

Figure 4.7. Graphical details of a typical simmerstat. (A, bimetal strip; B, load contacts; C, heater element; D, control spring; E, sprung metal delivery arm; F, adjusting pivot; G, cam; H, control knob.)

Figure 4.6 illustrates the internal switching and wiring arrangement of a typical simmerstat in schematic form whereas Figure 4.7 graphically dipicts the internal structure of the device.

Function

The role of the domestic simmerstat has been designed to prevent thermal over-run by a combination of manual control and automation and will monitor and regulate the temperature of a connected external heating load.

Inside the simmerstat there is a tiny heating element mounted on a small length of thermal insulation attached to a bimetal strip (Figure 4.8). Its function is to gently warm the bimetal component to effect linear movement. The heating element assembly is attached to a supporting cradle, lever arm and adjustable pivot which, when motivated thermally, provides movement for a snap

sprung-loaded switching action to occur. Once the load contacts are automatically opened, the small heating element loses its supply of electricity and cools. The reduction in temperature produces a displacement in the bimetal strip away from the mid-point of the simmerstat. This allows a reduction in pressure on the switching arm originated by the adjustable pivot attached to the

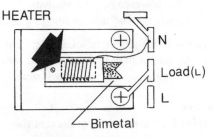

Figure 4.8. Inside the simmerstat there is a heating element (arrowed) mounted on a bimetal strip (see also 4.7(a)).

bimetal strip. When sufficiently cooled, the switching contacts will snap closed providing a path for current serving both external load and tiny integral heating element.

The procedure is then repeated.

Thermostats

Bimetal thermostat

Application

The role of a thermostat can be described as a thermomechanical device for maintaining a steady temperature.

One of the most common methods is by use of a thermo-sensitive bimetal strip adapted so that when heated or cooled by an *extraneous* source, movement is provided to automatically open or close electrical switching contacts.

The bimetal principle can be tested using two equal lengths of dissimilar metals such as *brass* and *iron* riveted together forming a single component. As both pieces of metal have different rates of expansion, the brass component will become longer than the iron when heated and cause the bimetal to bend as outlined in Figure 4.9. This represents the bimetal principle in its most basic form.

Thermostats operating on bimetallic expansion are employed for various applications which include the following:

1. Electrical controls; central heating systems.
2. Frost protection; controlling electrical trace heating.
3. Industrial heating arrangements.
4. Electric motor control, via a suitable contactor.
5. Portable heating appliance control.
6. Solar powered heating systems.
7. Ventilation and air-conditioning control.

Figure 4.10 illustrates in schematic form how a simple thermostat may be incorporated into the wiring serving a basic ventilation system.

Function

The role of a thermostat is to prevent *thermal over-run* by means of automatic control. Figure 4.11 schematically shows the internal wiring arrange-

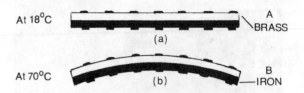

Figure 4.9. The principle of the bimetal strip (A, brass; B, iron): (a) before heat is applied; (b) after heat has been applied. The linear expansion of iron is 0.000012/K; the linear expansion of brass is 0.000019/K.

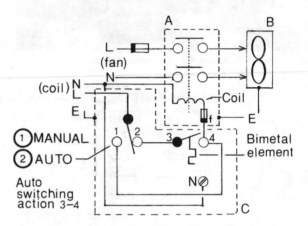

Figure 4.10. A thermostat used to control a basic ventilation system. (A, contractor; B, fan motor; C, thermostat.)

ments serving a domestic frost stat. When the connected load is less than 13 amps it is advisable to provide a 'neutral' conductor to terminal 4. This completes the *heat anticipator circuit*, the internal load taking the form of a small resistor located near the bend in the bimetal strip as shown in Figure 4.11. The resistor valued at approximately 0.4 megohms, gently preheats the bimetal strip to prevent over-run of the connected heating load. Some thermostats are provided with an adjustable heat anticipator circuit. This often proves to be beneficial when precision is demanded.

Operation

The working temperature is selected and set by an adjustable control knob attached by a spindle to a cam wheel. When a drop in ambient temperature is experienced the intergral bimetal strip involuntarily bends inwards as illustrated in Figure 4.11. A small permanent magnet fitted to the underside of the free-moving 'phase' contact provides a snap-on

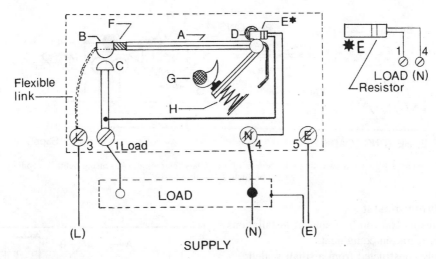

Figure 4.11. The internal wiring and switching arrangement serving a basic domestic thermostat. (A, flexible bimetal strip; B, flexible supply contact, often aided with a small permanent magnet; C, fixed load-contact; D, sprung heat sink; E, heat anticipator resistor, approximately 0.43 megohm; F, insulator; G, control spindle and cam; H, spring.)

action for the approaching load contacts. This mechanical principle operates well in reverse when a rise in ambient temperature is experienced and assists in preventing unnecessary arcing of the load contacts.

Working problems

Open type thermostats are fitted with ventilation grills which allow free movement of air over the active working parts. Although very efficient and easy to maintain, they often accumulate grime, dust, humidity and atmospheric pollutants that are, or have been, present. It is wise to thoroughly clean and service this type of thermostat from time to time, especially when used for agricultural purposes or if the device seems to have become inefficient. Once all extraneous debris is completely removed, clean the contacts by rubbing with a fine abrasive paper to clear accumulated dirt and pollutants. Check that all conductor terminations are secure and that the cabling is in a satisfactory condition.

Capillary thermostat

The *capillary* or *line voltage* thermostat is widely used where conditions are hostile or in situations where it would be impractical to employ a conventional temperature-sensing device, for example in an ice bank serving a bulk milk tank or high-temperature oven.

The capillary type thermostat is fully adjustable and is constructed enabling switching to be effective without the need for a supplementary contactor or relay to switch the load. Designed to suit various temperature ranges and current ratings, this type of thermostat is available in a choice of capillary lengths to suit practical needs.

Function and construction

Both capillary and sensing bulb are hollow and are made common to each other. At the manufacturing stage a small quantity of gas is added which expands and contracts proportionally to a rise or fall in ambient temperature. An increase in temperature gives rise to an increase in pressure within the sensing bulb and expands the bellows proportionally. On reaching the required working temperature a spring-loaded microswitch is mechanically activated by the bellows which either open or close the circuit they are controlling (Figure 4.12).

Capillary or line voltage thermostats can be used for a variety of applications:

- air-conditioning units
- commercial, domestic and industrial ovens
- cold rooms
- ice control in bulk milk tanks
- unit heaters (industrial warm-air fans)

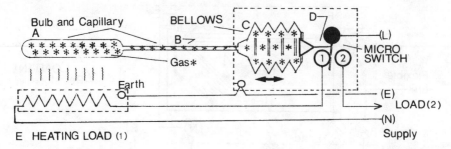

Figure 4.12. The internal wiring and switching arrangement of a capillary thermostat. (A, sensing bulb; B, capillary; C, bellows; D, microswitch; E, heating load.)

Mercury bulb thermostat

This type of thermostat can be found in installations controlling refrigeration equipment.

It is basically constructed from a small sealed glass or plastic tube in which a pool of mercury has been added. During the manufacturing process the mercury tube is attached to one end of a bimetal spiral arrangement while the other end is firmly anchored at the centre of the device (Figure 4.13). An increase or decrease in temperature will motivate the spiral pensing element to turn and in doing so will tilt the pool of mercury in the glass switching tube.

Switching action

Mercury switches are generally designed to provide for one- or two-way switching action as Figure 4.14 illustrates.

The two-way switch comprises a set of three post-contacts fitted within a sealed glass container in which a small pool of mercury has been placed. The *common* switching contact is ranged centre stage with the *normally open* and *normally closed* contacts flanked either side of it. When tilted anticlockwise the conductive pool of mercury makes a positive connection with the 'common' and 'normally closed' contacts, while reversing the movement bridges the 'common' and 'normally open' contacts. The density of mercury, 13.6 times greater than clean water, is of sufficient magnitude to provide instant contact thus minimising arcing when heavy loads are switched. The weight of mercury also counter-balances the action of the bimetal spiral and prevents any spring action from reversing the switching mode until a suitable temperature differential has been reached.

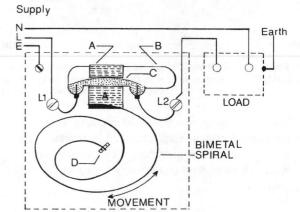

Figure 4.13. Typical mercury thermostat. (A, cradle; B, glass or plastic container; C, mercury; D, bimetal spiral centrally fixed.)

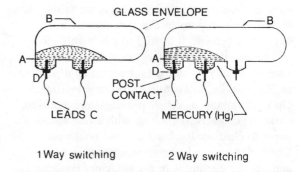

Figure 4.14. The mercury switch. (A, mercury; B, glass or plastic container; C, switching leads; D, post contacts.)

Advantages and disadvantages

Mercury switches are very reliable and will operate trouble free for many years. One advantage over other methods of temperature control is that the mercury in the sealed glass container is not subjected to contamination. When the pool of mercury is bright and shiny the switching action will function correctly. However, after years of use mercury often becomes dull and lifeless and when this happens the switch should be exchanged for an appropriate replacement.

Mercury bulb thermostats must be mounted horizontally to prevent switching problems. Manual control is effected by means of an external regulator designed to adjust the tension of a supplementary spring attached to the switching cradle.

Thermometers

Care must be taken not to confuse the concept of *temperature* with *heat energy* gained from it; they are completely different from each other.

The temperature of a body is a measurement of its *energy level* or 'hotness' and may be expressed in degrees Kelvin (K), Celsius (°C), Fahrenheit (°F) or Réaumur (°R). As an example, a large cylinder filled with warm water contains far more energy than the electric immersion element serving it.

Types in summary
The temperature of a body is measured by an instrument called a thermometer, of which there are many different types. Some are based on the expansion and contraction of liquids such as dyed alcohol or the liquid element mercury. Thermometers such as these are used for general and clinical purposes and in situations where temperatures are variable. Others are dependent on the movement of a spiral bimetal strip while work of a precise nature often requires a platinum resistance thermometer which converts the change in electrical resistance caused by differentials of temperature into degrees Celsius or Kelvin.

Thermoelectric thermometers are often used in industry in situations where other types would prove impractical. This variety of thermometer is formed by coupling two dissimilar metals such as *iron* and *constantan* (an alloy of copper containing between 10 and 55 per cent nickel), and recording the minute electrical current that is produced when heat is applied to it.

Lastly, the gas thermometer is often favoured in the refrigeration industry. This type functions by means of a change in gas pressure brought about by a corresponding change in temperature applied to a remotely positioned sensing phial.

To recap in listed form; the types of thermometer to be considered are as follows:

1. Clinical (liquid) thermometer.
2. Spiral bimetal thermometer.
3. Thermoelectric thermometer.
4. Gas thermometer.
5. Platinum resistance thermometer.

Consideration will be given to the first four, the platinum resistance thermometer being generally employed for work of a more specialised and precise nature.

Clinical thermometer
This specially designed instrument is used to measure the temperature of the human body. First aid boxes used on building sites are often supplied with one. Due to its special use and given that the temperature of a healthy human being is taken to be 37 °C (98.4 °F), the scale is abridged to read a few degrees either side of normal body temperature.

Construction
A clinical thermometer is constructed from a small elongated bulbous glass phial, in which mercury has been added and made common with a slender glass capillary. During the production stages the capillary is formed to a miniature 'humpback bridge' profile near to the end adjoining the phial. This deviant is known as a *constriction* and is illustrated as Figure 4.15.

Air is evacuated from the capillary and the capillary sealed. The complete assemblage is then enclosed within a moulded glass sleeve graduated in divisions of 0.1 °C from 35 °C to 43 °C as illustrated in Figure 4.16.

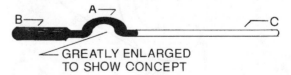

Figure 4.15. The constriction allows the body temperature to be read at an unhurried pace. (A, constriction; B, phial; C, fine capillary.)

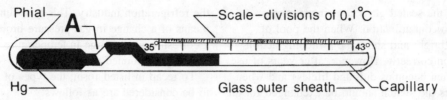

Figure 4.16. The clinical thermometer. (A, point of constriction.)

Working principle

An increase in temperature expands the mercury column through the area of constriction and along the capillary canal. Removing the thermometer from the patient's mouth produces a rapid cooling effect in the phial causing the column of mercury to retract breaking it at the point of constriction (Figure 4.16), leaving the leading column of mercury intact. The maximum temperature recorded can then be reviewed at an unhurried pace.

The divided column of mercury can then be rejoined by jolting or shaking the thermometer.

Spiral bimetal thermometer

This type of thermometer consists of a bimetal spiral-shaped cylinder constructed from two dissimilar metals such as *invar* and *brass* – invar

being an alloy, containing 0.2 per cent carbon, 36 per cent nickel and 63.8 per cent iron, which has a very low rate of linear expansion.

The outer end of the cylindical spiral is anchored as shown in Figure 4.17, leaving the other end free to move. A small spindle is attached to the unrestricted end on which a pointer is mounted with freedom to swing across a temperature scale. An increase in temperature influences the bimetal component and causes it to gyrate in a clockwise direction. As both pointer and spindle form an integral part of the spiral bimetal element, displacement resulting from thermal expansion will be magnified by the pointer as Figure 4.18 will illustrate. Similarly, a decrease in temperature will cause the bimetal to unwind or advance in an anticlockwise direction.

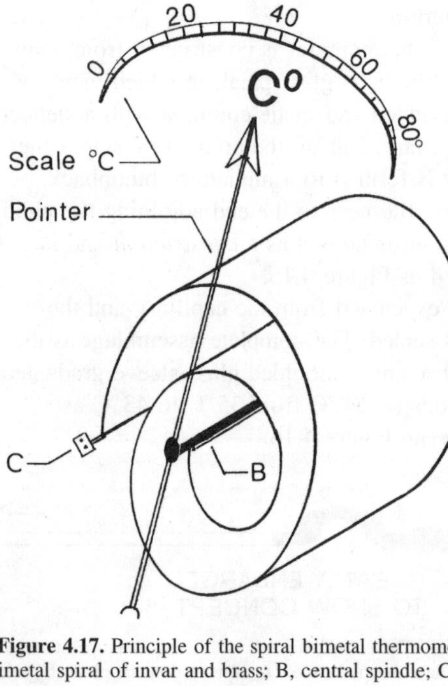

Thermoelectric thermometer

In 1821, *T.J. Seebeck* (1770–1831) discovered that a tiny electric current could be produced when two wires of equal length made from dissimilar metals, such as copper and iron, were twisted together at one end. He noticed that when heat was applied to

Figure 4.17. Principle of the spiral bimetal thermometer. (A, bimetal spiral of invar and brass; B, central spindle; C, secured point.)

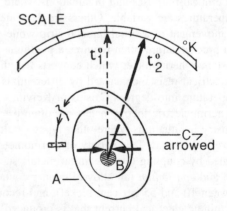

Figure 4.18. Displacement caused by thermal activity is magnified by the pointer. (A, bimetal element; B, spindle; C, thermal displacement.)

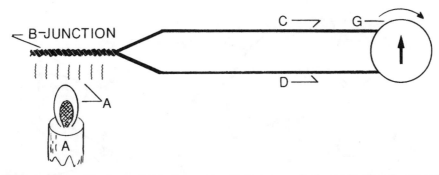

Figure 4.19. The principle of the thermoelectric thermometer. (A, heat source; B, thermal junction; C, copper; D, iron; G, galvanometer.)

the junction while a temperature differential was maintained at the other end, a minute current could be recorded when the two unattached ends were connected to a sensitive galvanometer (Figure 4.19).

This concept has been advanced over the years to produce the *thermopile*; an instrument used to measure temperature.

Construction

Commercially the instrument consists of a series of two dissimilar metals such as *antimony* and *bismuth* which are physically bonded at common junctions but insulated from each other, as illustrated in Figure 4.20. The two free ends are provided with high-temperature insulated leads which are connected to a sensitive galvanometer graduated in degrees Celsius or Kelvin.

The complete assemblage is packaged into a

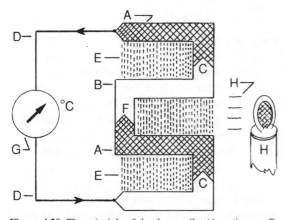

Figure 4.20. The principle of the thermopile. (A, antimony; B, bismuth; C, hot junction; D, insulated leads; E, insulation; F, cold junction; G, galvanometer; H, heat source.)

brass container in which a suitable sealer is added in order to protect and stabilise the working components.

Figure 4.20 depicts just two pairs of thermocouples but in practice a thermopile uses between 60 and 80 thermocouples all bonded to each other in a series formation.

This type of thermometer may be found in heavy industry where the need for such a device outweighs conventional methods of monitoring temperature variations. The thermopile may also be found in laboratories in the form of a portable or hand-held instrument.

Gas thermometer

Gases such as oxygen, hydrogen and nitrogen expand with reliable uniformity and make excellent thermometric 'fluids'. Mass for mass they expand far more than liquids subjected to the same temperature and are used in the construction of temperature-monitoring equipment supplied to the refrigeration and air-conditioning industries.

Structure and working principle

The gas thermometer consists of a small cylindrical lightweight metal reservoir some 150 mm long by 11 mm wide in which gas has been added. The reservoir is connected and made common with a long capillary tube which is coupled to a flexible sensitive hollow copper loop housed in the body of the thermometer (Figure 4.21(a) and (b)).

Variations in pressure, caused by a rise or fall in temperature, are transferred through the capillary to the copper-sensing loop causing a slight movement to occur. The end of the sensing loop is connected

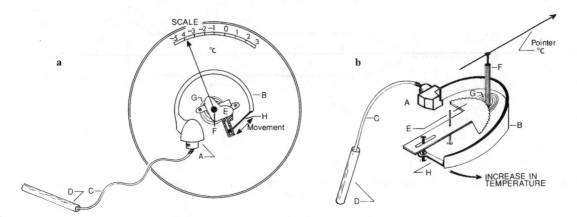

Figure 4.21. The principle of the gas thermometer. (A, pressure junction; B, copper-sensing loop; C, capillary; D, reservoir; E, ratchet; F, cog wheel spindle; G, hair spring; H, lever.)

to an arrangement of levers, cog-spindles and a semicircular ratchet to transfer thermolinear movement into an amplified semicircular motion enabling temperature values to be read. This is illustrated pictorially as Figure 4.21(b). The gas thermometer has many practical applications throughout the refrigeration industry.

This type of thermometer is very reliable; however, from time to time a small quantity of penetrating oil might be required to lubricate the moving components. Dryness and atmospheric pollution often cause the temperature indicator to stick, giving unreliable values.

Constant volume gas thermometer

This type of gas thermometer is extremely accurate and is ideally suited for laboratory work and thermo-standardisation projects.

Briefly, the constant volume gas thermometer comprises a gas bulb in which air has been removed and nitrogen added. The bulb is made common to a vertically positioned cylindrical glass mercury reservoir by way of a capillary tube (Figure 4.22).

Working principle

Variations in temperature permit the nitrogen gas to increase or decrease in volume allowing the mercury column to rise or fall in harmony with both pressure and temperature. As the constant volume gas thermometer is not very portable and requires considerable adjustment before use, it is not considered suitable for on-site applications.

Additional information may be gained from a more advanced textbook. The description and working concept of this special version of gas thermometer has been included to provide a brief insight and practical awareness of the types of thermometers available.

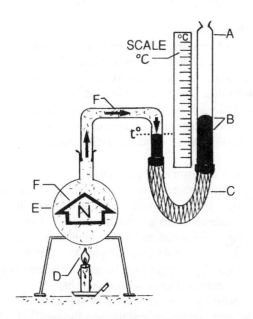

Figure 4.22. The constant volume gas thermometer. (A, reservoir; B, mercury; C, rubber connecting tube; D, heat source; E, gas bulb containing nitrogen gas; F, pressure.)

Linear expansion

Expansion can be the direct cause of many problems to an electrical installation. Hot weather or even warm ambient conditions can cause plastic (PVCu) conduit and trunking to warp or buckle under the stress caused by expansion.

Fortunately this can be easily avoided as a well-planned installation would demand that expansion couplers are used to reduce problems of this nature (Figure 4.23).

Advantages of linear expansion

Expansion also has its advantages. Hot rivets burred into shape will bring two work pieces tightly together when cooled. Thermometers rely on the expansion and contraction of liquids, such as mercury and alcohol, in order that temperature values may be read. Bimetal strips used in electrical thermostats and automatic flashing units fitted to directional indicators on cars both rely on thermal expansion and contraction in order to function.

Rates of expansion

All materials have different rates of linear expansion for each degree of temperature raised. Table 4.1 summarises in listed form several of the more common materials and their respective linear expansion rates.

The increase in length of a given substance per degree Celsius or Kelvin is known as the *coefficient of linear expansion* and is expressed as a factor or multiplier for a given substance, symbol α (lowercase Greek letter *alpha*).

To forward this concept: The coefficient of linear expansion for aluminium is 0.000 026/K, so every centimetre of a given length of aluminium, raised in

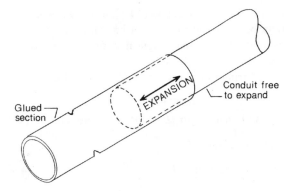

Figure 4.23. An expansion coupler.

temperature by 1 K, expands by 0.000 026 centimetre. To make this clearer: every kilometre of a given length of aluminium, raised in temperature by 1 °C, expands 0.000 026 kilometre.

Thus the coefficient of linear expansion, α, for a given substance can be defined by use of the following expression:

$$\alpha = \frac{\text{Increase in length}}{\text{Original length} \times \text{Rise in temperature}} \quad [4.1]$$

As an example, consider the following:

An aluminium conductor 200 cm long has a measured temperature of 20 °C. The conductor is placed in an industrial oven operating at a constant temperature of 100 °C. After a period of one hour the conductor is removed from the oven and remeasured with a laboratory micrometer and is found to be 200.416 cm in length. From the given data calculate the coefficient of linear expansion for aluminium.

Solution:
Referring to Expression [4.1] and the given data:
Increase in length = 200.416 cm − 200 cm
Original length   = 200 cm
Rise in temperature = 100 °C − 20 °C

$$\alpha = \frac{200.416 - 200}{200 \times (100 - 20)}$$

$$= \frac{0.416}{16\,000}$$

$$= 0.000\,026/°C$$

TABLE 4.1 A comparison of temperature coefficients for different materials per degree Celsius or Kelvin.

Compound or element	Temperature coefficient per °C or K
Aluminium	0.000 026
Brass	0.000 018 9
Copper	0.000 016 7
Glass (standard)	0.000 009
Invar alloy	0.000 000 9
Iron	0.000 012
Platinum	0.000 008 9
Silica	0.000 000 42

By transposing Expression [4.1], unknown values may be evaluated in terms of known data. For example, the original length may be determined in respect to other values given.

Referring to Expression [4.1]:

$$\alpha = \frac{\text{Increase in length } (L_i)}{\text{Original length } (L_o) \times \text{Rise in temperature } (^{\circ}t)}$$

First cross multiply,

$$\alpha \times L_o \times {^{\circ}t} = L_i \qquad [4.2]$$

Next divide each side by $\alpha \times {^{\circ}t}$

$$\text{Original length } L_o = \frac{\text{Increase in length}(L_i)}{\alpha \times \text{Rise in temperature } (^{\circ}t)}$$

$$[4.3]$$

All substances expand at different rates, and at times this can be beneficial. However, expansion can cause problems in electrical installation work. We must be aware of this in order to provide adequate safeguards to combat and control these problems.

Principle of moments

SI units and symbols

The SI unit derived from *force* is the *newton* and is defined as *the force required to produce an acceleration of 1 metre per second in a mass of 1 kilogram*, symbol N and named after *Sir Isaac Newton* (1642–1727).

The moment of a force about a point is quantified in *newton metres*, symbol N m.

The moment of a force

The moment of a force about a given point may be described as the 'turning effect' of the force about that point.

It may also be stated that a moment is equal to the product of the force (*N*) and the perpendicular distance (*d*) from the line of action of the force to the fulcrum.

Hence,

Moment of a force about a point $= N \times d$ [4.4]

This may be further simplified illustratively by

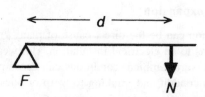

Figure 4.24. The moment of a force is equal to the product of the force, *N*, and the perpendicular distance from the force to the fulcrum, *F*.

studying Figure 4.24, where *F* is the fulcrum, *d* is the distance from the turning point to the fulcrum and *N* is the force applied in newtons.

When two or more forces are in a balanced equilibrium (Figure 4.25), the force applied in one direction must equal the force administered from the other direction. When two children of the same weight sit at either end of a play ground seesaw they will be seen to be a balanced equilibrium. This is known as the *principle of moments* and is defined as:

The sum of all clockwise moments =
The sum of all anticlockwise moments   [4.5]

Positive and negative moments

Traditionally, clockwise moments are annotated negatively while anticlockwise moments are positively annotated. Therefore, it may be said that the algebraic sum of all moments is zero. Take, as a simple example, Figure 4.25.

Anticlockwise moments: $+ (5 \times 2) = +10 \,\text{N m}$
Clockwise moments: $- (5 \times 2) = -10 \,\text{N m}$

Expression [4.5] can be confirmed by use of a strong metre rule of uniform mass and balanced at a mid-point position by means of a suitable fulcrum. Metric weights of known value should be

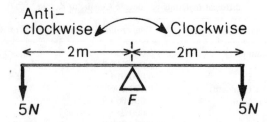

Figure 4.25. The sum of the clockwise moments equals the sum of the anticlockwise moments.

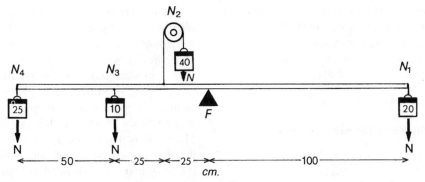

Figure 4.26. The principle of moments remains true no matter how many moments are operating at the same time.

suspended either side of the mid-point position with nylon thread and adjusted so that the metre rule is completely balanced.

A balanced equilibrium will observe the principle of moments no matter how many moments are operating at the same time, as Figure 4.26 clearly shows.

Clockwise moment (N_1):

$20 \times 100 = -2000$

Clockwise moment (N_2):

$40 \times 25 = -1000$

Total clockwise moments (N_1) and (N_2):

$-2000 + (-1000) = -3000$

Anticlockwise moment (N_3):

$10 \times 50 = +500$

Anticlockwise moment (N_4):

$25 \times 100 = +2500$

Total anticlockwise moments (N_3) and (N_4):

$+500 + (+2500) = +3000$

Algebraic sum of all moments: $3000 - 3000 = 0$

Finding the magnitude of an unknown force
Figure 4.27 illustrates a simple arrangement in which a uniform rod of negligible weight is balanced at its mid-point position by means of a suspended nylon thread. Two forces of known magnitude, (N_1) and (N_3), are suspended 2 metres from either side of the point of balance, F. A third and unknown weight, (N_2), is placed 1 metre left of

the fulcrum, F, and a balanced equilibrium is gained.

The mass of the unknown force may be determined by use of Expression [4.5] and referring to Figure 4.27.

$$(N_2 \times d_2) + (N_1 \times d_1) = N_3 \times d_3 \qquad [4.6]$$

$$N_2 \times d_2 = (N_3 \times d_3) - (N_1 \times d_1) \qquad [4.7]$$

$$N_2 = \frac{(N_3 \times d_3) - (N_1 \times d_1)}{d_2} \qquad [4.8]$$

Substituting, for known values:

$$N_2 = \frac{(100 \times 2) - (50 \times 2)}{1}$$

$$= \frac{200 - 100}{1}$$

$$= 100 \text{ N}$$

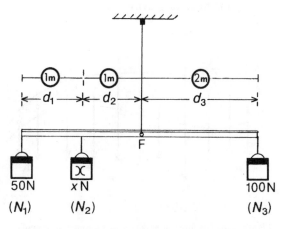

Figure 4.27. Finding the magnitude of an unknown force.

This expression may be rewritten to find the point of balance providing all values, other than one distance, are available. For example, d_3 may be expressed in terms of N_1, N_2, N_3 and d_2.

Referring to Expression [4.6]:

$$N_3 \times d_3 = (N_2 \times d_2) + (N_1 \times d_1)$$

Dividing each side of the equation by N_3

$$d_3 = \frac{(N_2 \times d_2) + (N_1 \times d_1)}{N_3} \qquad [4.9]$$

Substituting, for known values:

$$d_3 = \frac{(100 \times 1) + (50 \times 2)}{100}$$

$$= \frac{100 + 100}{100}$$

$$= 2 \text{ metres}$$

When a problem has to be solved involving several moments acting it can be an advantage to sketch and label the problem with all available data. Presented in this way the task will appear to be far clearer and easier to resolve.

Centre of gravity

Figure 4.28 illustrates in pictorial form an irregularly shaped sheet of glass.

All matter consists of a vast number of tiny molecules each possessing its own individual weight factor and each attracted to the centre of the Earth by the force of gravity.

(*A molecule is the smallest portion into which a substance may be divided, which is capable of an independent existence, yet still maintaining the properties of the original substance.*)

For practical purposes it is assumed that all forces acting upon the tiny molecules are both parallel and vertical to each other as the pull exerted by the Earth is such a vast distance away. The result is that the total weight of the body acts through a single point coinciding with the body's *centre of mass*, as illustrated in Figure 4.28. It is at this point that an irregular shape can be placed in a balanced equilibrium (Figure 4.29) and the centre of gravity located.

The centre of gravity of a body may be defined as a fixed point at which the sum of all molecular weights acts and where the force of gravity always passes.

Practical methods of locating the centre of gravity

There are two well-established methods of locating the centre of gravity of a manageable flat body such as a sheet of electrical insulating material or a small steel side panel serving a site-built assembly, and may be summarised as:

1. *Balancing method*
2. *Plumbline method*

Balancing method

A straight section of bevelled skirting board would be ideal when determining the centre of gravity of a workpiece, using this method.

(a) Fix the bevelled skirting board 90° to the finished floor.

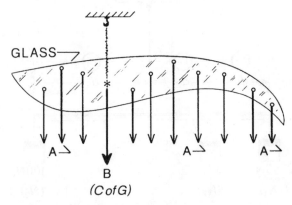

Figure 4.28. The force of gravity acting through an irregular-shaped sheet of glass. (A, the force of gravity acting on individual molecules; B, centre of gravity.)

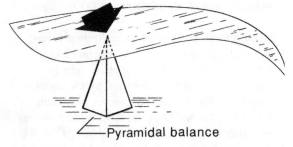

Figure 4.29. The centre of gravity may be found by placing the shape in a balanced equilibrium.

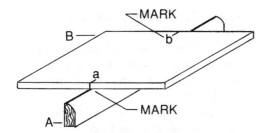

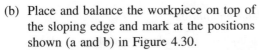

Figure 4.30. An alternative method of finding the centre of gravity. Mark a–b. (A, skirting board; B, workpiece.)

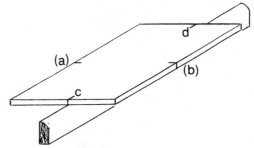

Figure 4.31. Rotate the workpiece through 90° and mark again, c–d.

(b) Place and balance the workpiece on top of the sloping edge and mark at the positions shown (a and b) in Figure 4.30.

(c) Rotate the workpiece horizontally through 90°; balance and mark again (c–d in Figure 4.31).

(d) Repeat placing the workpiece diagonally (e–f), as a means of checking.

(e) Join the marked balanced points, a–b, c–d and e–f with a pencil line as depicted in Figure 4.32.

(f) The centre of gravity will lie at the point where all three lines intersect.

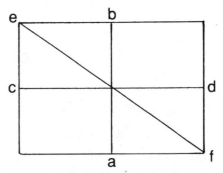

Figure 4.32. Join the marked balance points a–b, c–d and e–f with a pencil line.

Plumbline method

This procedure is ideal, if for example the centre of gravity has to be found for an irregularly shaped cardboard template used for a special electrical installation project.

(a) First punch a series of three or four suitably sized holes around the edge of the shape as described in Figure 4.33.

(b) Hammer a small masonry nail in a suitable timber overhang and place the template over the nail at hole A1 allowing it to swing freely.

(c) Once the template is at rest its centre of gravity will be at a point directly below the masonry nail.

(d) Carefully hook the end of the string of the plumbline onto the nail allowing it to hang freely without brushing the workpiece.

(e) Mark the template carefully at the top and bottom at a point where the plumbline intersects the edge of the template, as illustrated in Figure 4.34

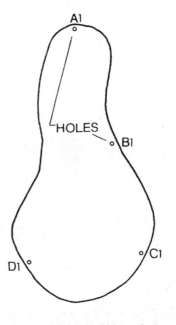

Figure 4.33. A series of holes should be made around the edge of the shape, A1, B1, C1 and D1.

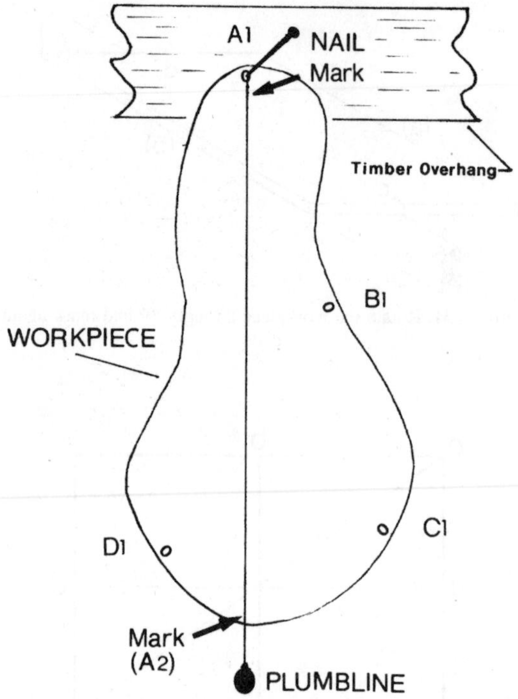

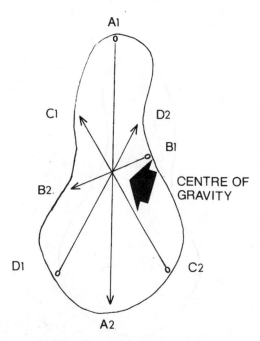

Figure 4.34. Mark the template carefully where the plumbline passes the edge of the template.

Figure 4.35. The centre of gravity lies at the point where all lines intersect.

(f) Repeat the test from positions B1, C1 and D1.

(g) Remove the template and place on a flat surface. Join together reference points A1–A2, B1–B2, C1–C2 and D1–D2 with a pencil line.

(h) The centre of gravity will lie at the point where all four lines intersect, as demonstrated in Figure 4.35.

Conditions of equilibrium

Stable equilibrium

If a body such as a factory-made steel electrical cabinet is slightly displaced, released, then returned to its original position, the centre of gravity *rises* during the period of displacement as illustrated in Figure 4.36. This condition is known as *stable equilibrium*.

Unstable equilibrium

If a body is slightly displaced and then released and the centre of gravity falls during the period of displacement, the condition is known as *unstable equilibrium*. Figure 4.37 illustrates this concept using a pyramid-shaped body balanced on its tip.

Neutral equilibrium

If a body is slightly displaced but settles in the same position when released, the centre of gravity will remain at a constant height during the period of displacement. This condition is known as *neutral equilibrium* and may be illustrated using a sphere, as shown in Figure 4.38.

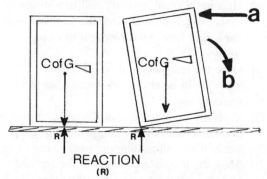

Figure 4.36. If a body is displaced (a), released (b), and returns to its original position, it is in a condition of stable equilibrium.

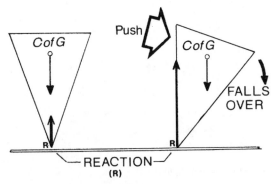

Figure 4.37. Unstable equilibrium.

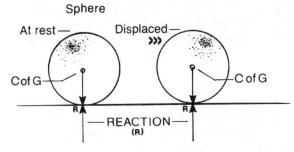

Figure 4.38. Neutral equilibrium.

Principle of levers

A lever can be defined as a rigid steel bar designed to turn freely about a fixed point, as illustrated in Figure 4.39(a). It can be used as an aid to displace a heavy load, such as a large electric motor, with comparative ease where manhandling would require the assistance of others.

Applied properly, a lever will provide an excellent mechanical advantage for the user. From experience it is far easier to displace a heavy load using a lever with a long *effort arm* and a short *load arm* than the other way round.

Mechanical advantage

The mechanical advantage, MA, of a lever can be defined as: *the ratio of the length arm from the fulcrum to the effort to the length of arm from the fulcrum to the load*. Hence,

$$MA = \frac{\text{Arm length from fulcrum to effort}}{\text{Arm length from fulcrum to load}} \quad [4.10]$$

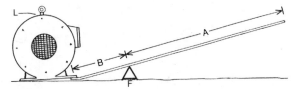

Figure 4.39(a). The principle of a lever. (A, effort arm; B, load arm; F, fulcrum; L, load.)

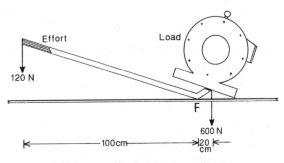

Figure 4.39(b). A lever is used to move a heavy electric motor.

It may also be evaluated by applying the principle of moments in relation to the load and effort.

The moment of a force about a point has been established in previous paragraphs as the product of the force and the perpendicular distance from the fulcrum. As an example, and referring to Figure 4.39(b):

$$\text{Load} \times 20 = \text{Effort} \times 100$$

Dividing each side of the equation by 'Effort × 20':

$$\frac{\text{Load}}{\text{Effort}} = \frac{100}{20}$$

Substituting for known values:

$$\frac{600}{120} = \frac{100}{20}$$

Therefore, MA = 5.

The *mechanical advantage* of a lever can be increased by employing a longer effort arm and this would have the effect of making the task easier.

Classes of lever

It might appear at first sight that all types of levers are governed by the same mechanical principles.

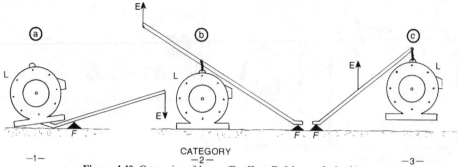

Figure 4.40. Categories of lever. (E, effort; F, fulcrum; L, load.)

This is not, so, as Figure 4.40 clearly demonstrates. Levers are divided and classified into three categories; each category applying different operational principles.

Category 1

Levers in this category operate when the fulcrum is placed between the load and effort as illustrated in Figure 4.40(a). Examples may be drawn from applying a crowbar to a heavy load or while trimming a conductor using a pair of insulated wire cutters.

Category 2

The second category of lever operates by positioning the fulcrum at, or towards, the end of the lever with the load situated between the fulcrum and effort (Figure 4.40(b)). Examples of use for this type of lever include removing a cap with a bottle opener, pushing a loaded wheelbarrow and fracturing nuts with a pair of nut crackers.

Category 3

The last category of lever has a mechanical *disadvantage* over other types as the fulcrum appears near or at the end of the lever while the effort is concentrated at a midpoint position between the load and the fulcrum (Figure 4.40(c)). An example of this type of lever can be witnessed when observing a fisherman land his catch using a conventional rod and tackle; the base of the rod being firmly supported against the angler's hip. Twin formations of this category of lever can be demonstrated when applying first aid tweezers to a wood splinter.

Defining work done

Work is said to be done when a force produces movement in the direction of the force. Whenever a diesel locomotive pulls a freight train, work is done; and also when an electrician climbs a ladder in order to carry out a task.

SI units

The SI derived unit of work is known as the *joule* (J), which may be defined as *the work done when a force of 1 newton (N), moves through a distance of 1 metre in the direction of the force.*

Larger units used are:

- the kilojoule (kJ) = 1000 J
- the megajoule (MJ) = 1 000 000 J

The unit of work was named after *James Prescott Joule* (1818–89).

Work is said to be done when a force produces movement in the direction of the force, therefore:

$$\text{Work} = \text{Force} \times \text{Distance moved in the direction of the force} \qquad [4.11]$$

or

$$W = F \times S \qquad [4.12]$$

where W is the work done in *joules* (J)
          F is the force applied in *newtons* (N)
and S is the distance moved in *metres* (m)

Examples of *work done* can be drawn from everyday life. Consider the following problem:

Calculate the work done in kilojoules by a small diesel forklift truck exerting a force of 1000 newtons over a distance of 80 metres.

Solution:

Referring to Expression [4.12]

> Work done = $F \times S$

Substituting for known values:

> $W = 1000 \times 80$
> $ = 80\,000\,\text{J}$

To convert to kilojoules, divide the answer by 1000. Hence,

$$kJ = \frac{80\,000}{1000}$$

Therefore,

> Work done over a distance of 80 metres = $80\,\text{kJ}$

Mass, weight and force

Mass

The mass of a body is the quantity of matter it contains and is unaffected by changes of gravity. A reel of insulated cable of a mass of 5 kg will maintain the same mass whether used on Earth or taken to the Moon. The mass of any given object will depend on its *density* (ρ) expressed in kilograms per cubic metre. A column of mercury would have a mass approximately 13.6 times greater than a similar column of water at 4 °C as Figure 4.41 illustrates. The mass of an object (m) can also be defined as the product of the density (ρ) of a substance in kilograms per cubic metre and the total volume (V) the substance occupies in cubic metres. Hence,

$$m = \rho \times V \qquad [4.13]$$

Mass is always calculated using a beam or top-pan balance and is never measured with a conventional spring balance which relies on the gravitational pull

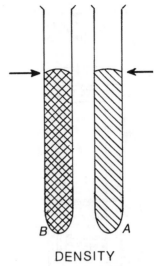

DENSITY

(Hg)13.6   (H₂O)1.0

Figure 4.41. A column of mercury has a mass 13.6 times greater than a similar column of water at 4 °C. (A, water; B, mercury.)

of the Earth in order that a measurement may be taken.

The basic SI unit for mass is the kilogram. This can be used in various forms of multiples and submultiples, as Table 4.2 will show.

Weight

The weight of an object is determined by the force of gravity and is measured in *newtons* (N).

Occasionally the weight of a body is expressed in units of mass. This is not technically correct and it is far more appropriate to define weight in terms of *newtons*. The weight of a body varies depending on where it is measured. An electrical accessory weighing 1 N on the surface of the Earth would, when weighed on the Moon, register a mere 0.166 N whereas taken on a journey to the giant gas planet Jupiter, would weigh a staggering 2.64 N.

In summary, the weight of a body can be defined

TABLE 4.2   SI units for mass

Unit	Symbol	Subunit and equal to	Symbol
1 metric tonne	t	1000 kilograms	kg
1 kilogram	kg	1000 grams	g
1 gram	g	1000 milligrams	mg
1 milligram	mg	1000 micrograms	μg

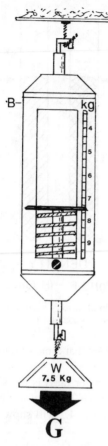

Figure 4.42. The weight of a body is defined as the force in *newtons* acting on a mass upholding it. (B, spring balance; G, force of gravity; W, weight.)

as the force in *newtons* acting on a mass upholding it, as Figure 4.42 clearly illustrates.

Force

Force can be described as an extraneous influence which alters or tends to change the state of rest or motion in a body and is measured in *newtons*. A *newton* can be considered approximately equal to the weight of a 100 g mass but formally is defined as *the force needed to produce an acceleration of 1 metre per second² in a mass of 1 kg.*

Expressed in newtons

The force of gravity together with the rotation of the Earth on its axis determines the weight of a body.

The acceleration experienced by a free-falling

mass of 1 kg placed in a vacuum and tested at various places has been evaluated to average 9.8 metres per second². The invisible force acting on this unit mass must be 9.8 newtons in order to produce an acceleration of 9.8 m/s² since a force of 1 newton produces an acceleration of 1 metre per second² on a mass of 1 kg.

With these figures fresh in mind, it may be stated that the *weight* of a 1 kg mass is 9.8 newtons.

Put another way, a body of mass m kg experiences a force of N newtons when allowed to fall freely within a gravitational field.

The value of the force (F) expressed in newtons (N), will be determined by the mass of the object and may be found by evaluating the product of the falling mass in kilograms and the gravitational constant g. Hence:

$$F = m \times g \qquad [4.14]$$

where g is 9.81 metres/second².

When a body, such as a large electric motor, is at rest, gravity is still acting upon it, for if this were untrue an unsecured motor would float. The force experienced due to gravity on an object at rest is called *weight* and is measured in *newtons*. Therefore it may be stated that:

$$\text{The weight (N) of a body} = m \times g \qquad [4.15]$$

As a practical example consider the following problem:

Calculate the work done in joules by an electrician of 70 kg mass climbing a vertical service ladder 7 metres high, given that the average value of 'g' is 9.81 m/s².

Solution:
Referring back to Expression [4.11]:

Work done in Joules = Force × Distance moved

Work done in Joules in opposing gravity =
$$(m \times g) \times d \qquad [4.16]$$

Substituting for known values:

$$\text{Work done} = (70 \times 9.81) \times 7$$
$$= 4806.9 \; J$$

Efficiency of a machine

A machine may be described as a device for overcoming physical resistance at one point by the application of a force at a point elsewhere. Wire cutters, crimping tools and pulley wheels can all be regarded as machines. Regrettably they will always operate at less than their full potential as they are in constant opposition to extraneous and internal losses such as gravity, air resistance and friction.

If it were possible to design a machine with an efficiency of 100 per cent, the useful work output would always equal the work input. Unfortunately, no matter how well conceived, output will never exceed input. At times this concept can appear confusing but mathematically it may be proved to be correct.

Expressed as a percentage

The efficiency of a machine (η) is usually expressed as a percentage and can be defined as the ratio of the useful work output to the applied work input.

Hence:

$$\text{Efficiency} = \frac{\text{Work done by load}}{\text{Work done by effort}} \times \frac{100}{1} \quad [4.17]$$

As an example, consider first a theoretically perfect machine taking the form of an electrical contractor's block and tackle, illustrated as Figure 4.43.

Referring back to Expression [4.11]:

Work done in joules
= Force in newtons
× Distance moved in metres

Work done by load = 100 × 2
= 200 J

Work done by effort = 25 × 8
= 200 J

$$\text{Efficiency} = \frac{\text{Work done by load}}{\text{Work done by effort}} \times \frac{100}{1} \quad [4.17]$$

$$\eta = \frac{200}{200} \times \frac{100}{1} = 100 \text{ per cent}$$

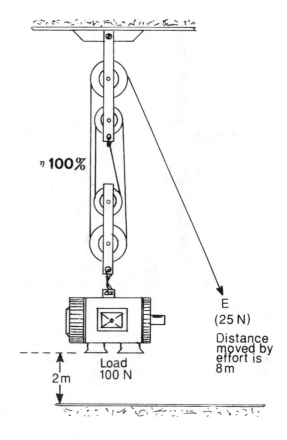

Figure 4.43. A theoretically perfect machine. (E, effort. Distance moved by effort, 8 metres. Distance moved by load, 2 metres.)

Unfortunately, in practical terms this is not possible as the following example will help to confirm:

Figure 4.44 depicts an electric motor weighing 100 newtons being raised through a vertical distance of 2 m by a poorly maintained block and tackle. The effort required to raise the load is found to be 40 newtons moved through a distance of 8 metres.

In order to calculate the efficiency of this machine, work done by both load and effort must first be evaluated:

Work done in Joules = Mass × Distance moved

Work done by load = 100 × 2 = 200 J

Work done by effect = 40 × 8 = 320 J

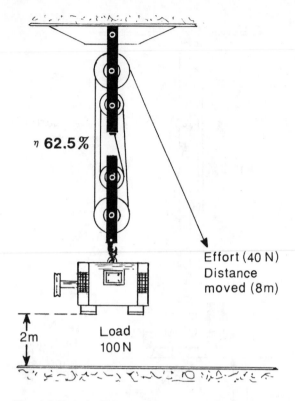

η **62.5%**

Effort (40 N)
Distance
moved (8m)

Load
100 N

2m

Figure 4.44. An inefficient machine with an efficiency of 62.5 per cent.

$$\text{Efficiency} = \frac{\text{Work done by load}}{\text{Work done by effort}} \times \frac{100}{1} \qquad [4.17]$$

$$\eta = \frac{200}{320} \times \frac{100}{1}$$

$$= 62.5 \text{ per cent}$$

This is clearly a very inefficient machine but machines with such low efficiency do exist wherever their maintenance is overlooked.

Velocity ratio and mechanical advantage

The *velocity ratio* (VR) of a machine is the ratio of the distance through which the applied effort moves to the distance moved by the load at the same time. When the applied effort is smaller than the load, the velocity ratio is always greater than unity, so heavy loads can be moved slowly with comparative ease by moving the effort briskly.

By observation the VR can be determined by the number of *effective strings* serving the lower arrangement of pulleys. The block and tackle illustrated as Figure 4.43 has a velocity ratio of 4. Alternatively, VR may be calculated by use of the following expression:

$$VR = \frac{\text{Distance moved by effort}}{\text{Distance moved by load}} \qquad [4.18]$$

Reflecting back to Figure 4.43:

$$VR = \frac{8}{2} = 4$$

Mechanical advantage

The *mechanical advantage* (MA) of a machine can be expressed as the ratio of the actual load raised to the effort required to raise the load and may be emphasised as:

$$MA = \frac{\text{Load raised in newtons}}{\text{Effort required in newtons}} \qquad [4.19]$$

Efficiency in terms of VA and MA

The efficiency of a machine (η) may also be evaluated in terms of the velocity ratio and mechanical advantage by use of the following expression:

$$\eta = \frac{MA}{VR} \times \frac{100}{1} \qquad [4.20]$$

Referring back to Figure 4.44, Expressions [4.19] and [4.18] and recapping:

- Load = 100 N
- Effort = 40 N
- Distance moved by load = 2 m
- Distance moved by effort = 8 m

Therefore,

$$MA = \frac{100}{40} = 2.5$$

$$VR = \frac{8}{2} = 4$$

Hence,

$$\text{Efficiency } (\eta) = \frac{MA}{VR} = \frac{2.5}{4} \times \frac{100}{1}$$

$$= 62.5 \text{ per cent}$$

Wasted energy

The percentage of energy wasted by a machine in overcoming gravity, friction and air resistance can be evaluated by subtracting the *efficiency factor* from 100 per cent. Hence:

$$\text{Wasted energy} = 100 - \eta \qquad [4.21]$$

Referring back to Figure 4.44:

$$\text{Wasted energy} = 100 - 62.5$$

$$= 37.5 \text{ per cent}$$

(In a perfect machine the velocity ratio is equal to the mechanical advantage.)

Efficiency of tungsten filament lamp

This may be expressed as:

$$\eta = \frac{\text{Power consumed in watts (W)}}{\text{Illuminating power in lux (lx)}} \qquad [4.22]$$

Efficiency of an electric motor

The efficiency of an electric motor will depend on the following criteria:

1. Iron losses.
2. Magnetic losses.
3. Gravity.
4. Friction.
5. Air resistance.

The efficiency of an electric motor is expressed as a percentage which may be calculated from the following expression:

$$\eta \text{ (Motor)} = \frac{\text{Power out}}{\text{Power in kilowatts}} \times \frac{100}{1} \qquad [4.23]$$

Use of block diagrams

Block diagrams are a useful way to forward a concept without the need for technical detail. They also provide means of focusing on potential problem areas and enable methodical planning to be carried out.

Figures 4.45 to 4.49 are illustrative examples of block diagrams required to be identified by students registered for the National Vocational Qualification in Electrical Installation Engineering.

Summary

1. The kilowatt-hour induction meter is a skeletal electric motor in which the armature is formed from a ferric disc.
2. A modern radio telemeter has incorporated a miniature FM receiver/switch which accepts signals from BBC Radio at predetermined times of the day. This effectively switches the restricted section of the telemeter to provide off-peak power for the consumer.
3. A simmerstat is a device used to regulate heating loads and prevent thermal over-run.
4. The role of a thermostat is to maintain a steady working temperature by means of thermomechanical switching.
5. Capillary, or line voltage, thermostats are used where conditions are hostile, as for example in an ice bank serving a bulk milk tank or in a domestic oven.
6. A mercury bulb thermostat is constructed from a small sealed glass or plastic tube in which a pool of mercury has been added together with two or three switching-post contacts. Often used to control refrigeration equipment and sump pumps.
7. Temperature is a measurement of a body's energy level or 'hotness' and is expressed in degrees Celsius or Kelvin.
8. The increase in length of a given substance per degree Celsius or Kelvin is known as the

Figure 4.45. Direct current power supply.

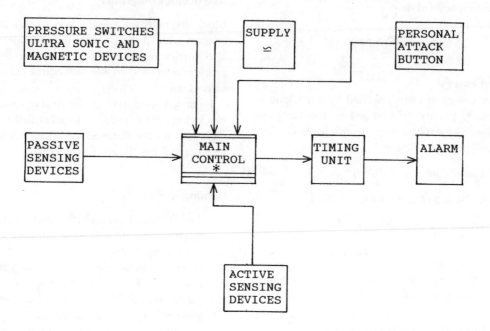

Figure 4.46. Basic security system.

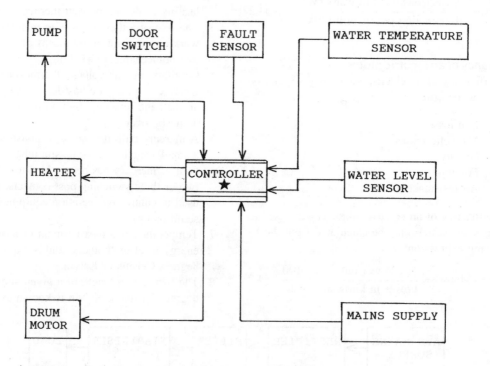

Figure 4.47. Washing machine control.

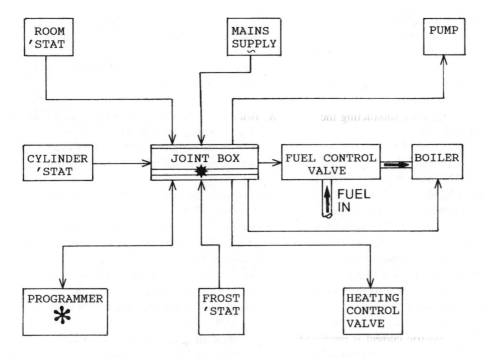

Figure 4.48. Space heating control.

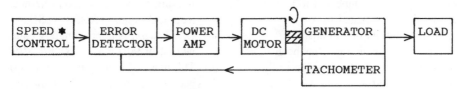

Figure 4.49. Motor speed control.

coefficient of linear expansion and is expressed as a numerical factor.

9. In a body experiencing a state of even balance, the total sum of all clockwise moments will equal the sum of all anticlockwise moments.

10. The centre of gravity of a body may be defined as a fixed point at which the resultant of all molecular weights acts and where the force of gravity always passes.

11. The centre of gravity of a body may be defined by either using the plumbline method or by balancing. The three states of equilibrium are: stable, unstable and neutral.

12. The mechanical advantage of a lever can be defined as the ratio of the length of arm from the fulcrum to the effort to the length of arm from the fulcrum to the load.

13. Work done is defined as when a force produces movement in the direction of the force. The SI derived unit of work is the *joule* (J). See also Appendix B.

14. The mass of a body is the quantity of matter it contains. It is completely unaffected by gravity. (Mass = Density × Volume.)

15. The weight of a body is determined by the force of gravity and is measured in *newtons* (N).

16. Force can be defined as the effort which influences or alters the state of rest or motion experienced in a body and is measured in *newtons*.

17. The efficiency of a machine is usually expressed as a percentage and it is the ratio of useful work output to applied work input.

18. The velocity ratio of a machine (VR) is the ratio of the distance moved by the applied effort to the distance moved by the load at the same time.
19. The percentage of energy wasted by a machine in overcoming friction and the effects of gravity may be evaluated by subtracting the efficiency factor from 100.
20. Clinical thermometers are of special design in order to accurately measure the temperature of the human body. The thermometer is graduated in divisions of 0.1 °C from 35 °C to 43 °C.
21. A spiral bimetal cylinder thermometer is constructed from a spiral cylinder made from two dissimilar metals such as *invar* and *brass*. A temperature differential influences the bimetal component and causes it to gyrate. A spindle and pointer are attached to the bimetal element of the thermometer.
22. A thermoelectric thermometer operates by the principle that a tiny electric current is produced when two wires of equal length and made from dissimilar metals are twisted together at one end. The two free ends are connected to a sensitive galvanometer graduated in °C.
23. A gas thermometer consists of a small cylindrical reservoir into which gas has been added. A capillary tube is made common to the reservoir and is terminated at a hollow sensing loop. Variations in pressure due to temperature changes cause the sensing loop to move away from itself. Temperature is indicated through a series of levers and cogspindles attached to a pointer.
24. A constant volume gas thermometer is used for laboratory and thermo-standardisation projects.
25. Different material substances expand at different rates when subjected to the same rise in temperature.

Review questions

1. List three types of kilowatt-hour meters.
2. Briefly describe the role of a simmerstat.
3. Suggest two metals that could be used to demonstrate the bimetal principle.
4. Describe a problem often affecting open type thermostats.

5. Relate to a practical problem in electrical installation engineering concerning linear expansion.
6. State briefly the principle of moments of a force about a point.
7. What is meant by the term *centre of gravity*?
8. Briefly describe how the *mechanical advantage* of a lever may be defined.
9. How is the weight of a body determined? State one method of calculating the weight of an object.
10. How may the velocity ratio of a machine be determined?
11. List the three conditions of equilibrium.
12. The mechanical advantage of a lever can be increased by
    (a) employing a longer load arm
    (b) employing a longer effort arm
    (c) increasing the pressure on the effort arm
    (d) increasing the size of the fulcrum.
13. Confirm the following statements:
    (a) Block diagrams are a useful way to forward a concept without the need for technical detail. TRUE/FALSE
    (b) One *newton* is approximately equal to the weight of 90 g mass. TRUE/FALSE
    (c) The mass of a body is never determined by use of a beam or top-pan balance. TRUE/FALSE
    (d) Category 3 levers have a mechanical advantage over other types of levers as the fulcrum appears near or at the end of the lever. TRUE/FALSE
14. List the two active components of a thermoelectric thermometer.
15. Itemise three types of thermometer reviewed in this chapter.

Handy hints

1. Tungsten halogen lamps will eventually blacken when controlled by a dimmer switch set to a reduced voltage.
2. Recommended temperature settings for thermostats:
    (a) hard water areas, up to 65.5 °C (150 °F)
    (b) soft water areas, up to 82.2 °C (180 °F)

3. The efficiency of a hot water immersion heater cylinder may, as a guide, be taken as:
   (a) unlagged: from 77 to 85 per cent
   (b) lagged: from 90 to 95 per cent
   (c) moulded lagged: from 95 to 98 per cent

4. The time required to bring a quantity of water stored in an immersion heater vessel to a predetermined temperature may be evaluated by use of the following expression.

$$\text{Hours} = \frac{\text{Litres of water} \times \text{Temperature rise (°C)}}{\text{Efficiency} \times \text{kW Rating of heater} \times 860}$$

[4.24]

(Efficiency must be expressed as a decimal using this expression; 1 *kg/calorie* = the heat to raise 1 litre of water; 1 kilowatt-hour = 860 calories.)

5. The *efficacy* of a lamp is the ratio of light output in lumens to the lamp power in watts.

6. Misalignment is often to blame for vibration in electric motors. Other reasons are:
   (a) unbalanced rotor due to careless servicing.
   (b) badly worn bearings.
   (c) loose motor mounting bolts.
   Check out a problem by a process of elimination when in doubt.

7. Equipment containing electronic components should not be tested using a proprietary portable appliance tester. The high voltages incurred when *insulation* or *flash* testing is carried out can seriously damage printed circuit boards.

5 Lighting

In this chapter: One-way switching and incandescent lighting. Two-way switching and tungsten halogen lighting. Intermediate switching and low-pressure mercury vapour lighting. Stroboscopic effects. High-pressure discharge lamps. Neon lighting. Regulations governing low-voltage discharge lighting. Extra-low-voltage lighting. Safe disposal and handling of lamps. Lamp bases. Application of lamps. The lumen. Alternative switching arrangements.

Lighting installations

Cables serving lighting installations must be chosen for their cross-sectional area with due consideration to whether they can withstand the conditions in which they are placed. For example, it would hardly be appropriate to plan for a street lighting installation using PVC-insulated and sheathed cable laid directly into the ground or to design a lighting scheme serving a boiler room wired in PVC singly insulated conductors drawn through PVCu conduit. In both cases problems would arise causing damage to the installation.

Consideration must be made concerning the number of lights wired in each circuit and the size of conductor used. Remember, also, that on very long runs of cable volt drop must be taken into account. Ideally a lighting circuit should have no more than ten individual lamps or a maximum current flow of 5 amps and wired using 1.0 or 1.5 mm^2 conductors. Over-current protection must match the potential design load of the circuit and the size of the conductor used.

Choice of lamps

Lighting schemes for large projects are now usually computer aided and are considered a highly specialised branch of electrical engineering. The effective design of modern lighting arrangements is beyond the scope of this book but types of luminaires and the methods of control available will be reviewed and examined highlighting both advantages and disadvantages of any such arrangement.

One-way switching and incandescent lighting

Modern ceiling roses are constructed to enable wiring and connecting to be carried out with the minimum of trouble, thus reducing the dependence of joint boxes.

Most ceiling roses are made with three independent sections, each electrically isolated from its adjacent neighbour, in which conductors are terminated (Figure 5.1). The centre section is traditionally reserved for the incoming and outgoing phase conductors while the neutral is terminated within the terminal block ranged to the left.

Once the supply cable has been terminated, a separate cable is routed from the ceiling point to serve a local control switch where the two current-carrying conductors and the current-protective conductor (CPC) are connected (Figure 5.2). If a PVC-insulated and sheathed cable served with

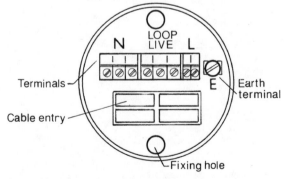

Figure 5.1. Wiring arrangements serving a typical ceiling rose.

Lampholder (L & N)

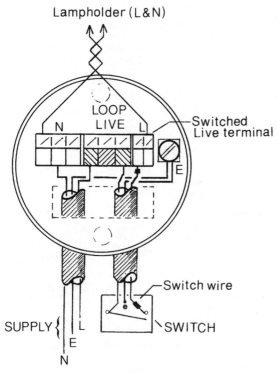

Figure 5.2. A cable is routed from the ceiling rose to serve a local control switch.

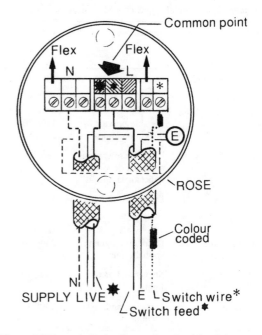

Figure 5.3. The switch wire is ranged to the right of the centre and made common with the flexible cable serving the lampholder. Remove only sufficient insulation in order to terminate within the ceiling rose.

traditional red and black coloured insulation is used, other than special purpose switching cable supporting two red insulated conductors, the black *switch wire* must be phase coloured at the ceiling point and switch. This provides means for identification and will alter the status of the black coloured conductor from a neutral to a *switched phase conductor*, commonly known as the *switch wire*. This may be carried out using coloured plastic oversleeving or PVC electrical insulation tape to BS 3924. Reference is made to the *Wiring Regulations, Table 51A*.

When terminating conductors at a ceiling rose the red core is made common with the incoming phase conductor and placed in the middle section. The black colour-coded switch wire is ranged to the right of the centre and made common with the flexible cable serving the lampholder, as illustrated in Figure 5.3. All protective conductors are sleeved using green/yellow plastic oversleeving and made common with each other within the termination point provided as required by Regulations 514–03–

01 and 514–06–02.

Figure 5.4 illustrates the basic circuit arrangement serving a one-way lighting arrangement. Graphical symbols used are to British Standard 3939.

Incandescent lamps

General-purpose electric lamps, as known today, were first produced in 1906 originating from the early carbon rod vacuum glow lamps experimentally produced in the late 1850s. Today we are spoilt for choice. Table 5.1 lists seven

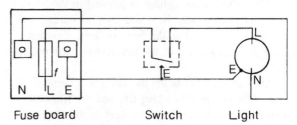

Fuse board   Switch   Light

Figure 5.4. A basic one-way switching arrangement.

TABLE 5.1   General-purpose incandescent (filament) lamps

Base abbreviations: BC, bayonet cap; SBC, small bayonet cap; ES, Edison screw; SES, small Edison screw; GES, Goliath Edison screw

Type	Application	Voltage	Wattage	Base	Average lumins per watt	Average life in hours
General-purpose lamp	Domestic	230/240	25–500	GES ES BC	10–15	1000
Rough service lamp	Portable or hand inspection lamps	110/120 220/250	40–100	BC ES	8.4	1000
Pygmy lamp	Used where space is restricted or as an indictaor lamp	230/240	15–25	BC SBC SES	7.3	1000
Candle lamps	Decorative lighting	230/240	25/60	BC SBC SES	8.6	1000
General-purpose coloured lamp	Decorative lighting, special effects	230/240	15–60	BC	–	1000
Striplight	Shaver and picture lights	230/240	30–60	S.15	6.6	1000
Night lights	Nurseries and night lighting	230/240	Very low	BC	–	1000

familiar types which are well known to the electrical industry, together with relative details.

Construction

General-purpose lamps are assembled using a glass bulb or elongated tube filled with a pressurised mixture of *nitrogen* and *argon* gases. This allows the lamp to burn at a much higher temperature and helps to prevent the tungsten evaporating from the surface of the coiled filament. The bulb is factory sealed to an integrally fitted glass support designed to carry the filament. In order to enable better efficiency of the lamp and reduce heat loss, the filament is coiled and reinforced by supplementary supports attached to the main glass pillar (Figure 5.5).

When the correct voltage is applied to the filament via the lamp base and integral wires, energy is dissipated in the form of heat, increasing until the temperature of the filament is sufficient to provide maximum light output.

After about 1000 hours the lamp will come to the end of its designed working life and the filament will rupture. This can cause a very high current to flow for a fraction of a second; often sufficient to

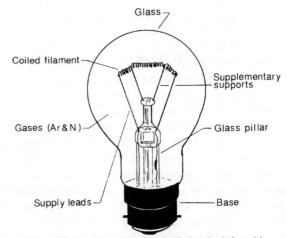

Figure 5.5. The lamp filament is coiled and reinforced by supports to enable better efficiency and reduce heat loss.

throw out a 5 or 6 amp miniature circuit breaker protecting the lighting circuit. High-wattage lamps, however, are provided with a tiny integral fuse within the body of the lamp to prevent damage occurring at the end of the lamp life.

Bases

Figure 5.6 depicts bases conforming to IEC 61–1

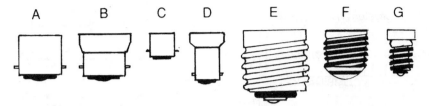

Figure 5.6. Lamp bases conforming to IEC 61-1 and BS 5101. (A, bayonet cap; B, bayonet cap skirted; C, small bayonet cap; D, small bayonet cap skirted; E, Goliath or giant Edison screw; F, Edison screw; G, small Edison screw.)

which are common to all incandescent lamps. The *Goliath Edision screw* (GES) design is reserved for lamps of a higher wattage such as 300 and 500 lamps used for floodlighting purposes.

Ageing process

During the life of a lamp, tungsten is being constantly evaporated from the filament, eventually causing the bulb to blacken and reduce light output. British Standards require that a lamp should not drop in illumination any lower than 85 per cent of its initial output level after a period of the first 750 hours.

Two-way switching and tungsten halogen lighting

Figure 5.7 illustrates how a two-way switching arrangement may be wired using the *joint box method*. In practice the circuitry is arranged using two- and three-core PVC-insulated and sheathed cable when applied to a domestic installation,

whereas single PVC-insulated conductors drawn through PVCu or steel conduit are more suitable for industrial and commercial purposes. Alternatively, the same switching arrangement may be wired without the use of a joint box and is known as the *conversion method*. This has the advantage of placing all terminations on one level making jointed conductors completely accessible. This method is ideal when it is impractical to install joint boxes within a ceiling void, if, for example, tiles are to be laid on the floor above. Many electricians find that this method is far easier.

Figure 5.8 explains how the conversion method of wiring a two-way switching circuit may be implemented.

Construction

The *Osram Company* were first to introduce halogen lamps into Europe in 1961, thus making a major contribution to both commercial and industrial lighting arrangements.

The lamp is fashioned from a single tungsten coiled filament sealed within a quartz glass tube or bulb in which a high-pressure inert gas has been added together with a measured amount of a halogen such as *bromide* or *iodine*. (*An inert gas can be described as being chemically inactive and*

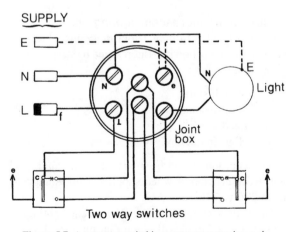

Figure 5.7. A two-way switching arrangement using a six-terminal joint box.

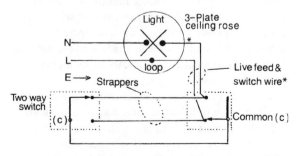

Figure 5.8. An alternative method of wiring a two-way switch.

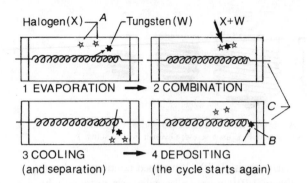

Figure 5.9. The operational cycle of a typical tungsten halogen lamp. (A, halogen; B, tungsten; C, filament leads.)

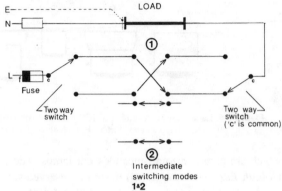

Figure 5.10. Intermediate switching arrangement.

may be one of the following gaseous elements: argon, helium, krypton, neon, radon or xenon.)

As the filament is designed to operate at very high temperatures it is not advisable to handle the glass mantle with bare fingers. This will avoid grease deposits from contaminating the wall of the lamp causing hot spots to occur when illuminated. Once a hot spot is established it does not take long before a small hole is blown from the side of the lamp.

Operation

To a lesser extent, compared with an incandescent lamp, tungsten is being constantly evaporated from the surface of the filament. Were it not for the higher pressure involved, the rate of evaporation would be similar.

Once evaporated (Figure 5.9) the tungsten combines chemically with the halogen content to form a *tungsten halide gaseous compound*. Providing the quartz glass wall is maintained at a constant high temperature the gaseous tungsten will not condense on the sides of the lamp and 'blackening' will be avoided. Lamps such as these are not suitable for dimming circuits or when used with open-fronted fittings in environmentally extreme cold conditions which might cause the wall of the bulb to reduce significantly in temperature to allow blackening to occur.

The hot compound of evaporated tungsten and halogen gas will follow natural convection streams around the interior of the glass envelope, and when returning to the filament will surrender the tungsten. The cycle continues indefinitely.

Unfortunately for the customer, the tungsten is not always returned to the same spot where evaporation took place otherwise the lamp could last for ever!

Intermediate switching and low-pressure mercury vapour lighting

Figure 5.10 illustrates how an intermediate switch may be added to or combined with, a two-way switching arrangement serving a low-pressure mercury vapour luminaire, better known as a fluorescent.

This method of switching employs one pair of two-way switches but there are no numerical constraints as to the number of intermediate switches which may be included within the switching arrangement.

Control gear: fluorescent lighting

Figures 5.11 to 5.14 depict four diagrams which are typical for the following fluorescent lighting control gear circuits.

1. Switch start
2. Semi-resonant
3. Lead-lag
4. High frequency.

Fluorescent lamps can be mounted at any angle and will work efficiently when supplied with voltages from 226 to 276 volts AC.

Other than high-frequency fluorescent lighting, most manufacturers will claim that the useful service life of a lamp is approximately 8000 hours

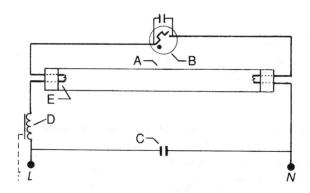

Figure 5.11. Fluorescent lighting control gear: switch start operation. (A, lamp; B, starter; C, power factor capacitor; D, ballast unit; E, heaters.)

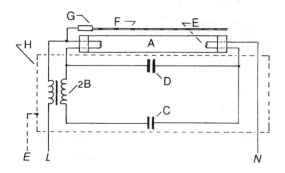

Figure 5.12. Fluorescent lighting control gear: semi-resonant circuit. (A, lamp; 2B, double ballast unit; C, capacitor; D, radio interference capacitor; E, heaters; F, external starting aid strip; G, high-value resistor built into the lamp base; H, ballast unit enclosure.)

based on a 3 hour switching cycle over a 24 hour period.

The fluorescent lamp

The vast majority of light emitted by a fluorescent lamp is caused by ultravoilet light bombarding and exciting a phosphorous coating on the inside of the lamp. Unlike an incandescent lamp (*filament lamp*), a fluorescent lamp does not rely on electrically heating a tungsten filament to a high temperature in order to produce visible light. A typical fluorescent lamp is manufactured from a long glass tube in which tungsten-coated cathode emitters have been fitted to each end (Figure 5.15). The lamp is supplied with a low-pressure inert gas such as *krypton* or *argon* and a small amount of the liquid element *mercury* while the glass interior is sprayed

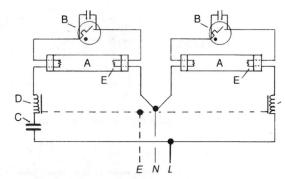

Figure 5.13. Fluorescent lighting control gear: lead–lag. (A, lamp; B, starter; C, series capacitor; D, ballast unit; E, heaters.)

with a coating of phosphorus.

When the fluorescent lamp is energised, both series connected cathode emitters are switched on for a brief period of time by means of an integrally fitted starter switch. This will ensure that there are sufficient free electrons flowing from the surface of the emitter. After a few seconds the heaters are automatically switched off and replaced by a supply voltage between the two ends of the lamp. Unfortunately for the user, a small amount of the cathode is consumed each time the light is switched on, so regrettably the lamp will experience a shorter useful life if constantly switched on and off throughout the day. Most makers of fluorescent lamps will quote a switching cycle of eight over a period of 24 hours.

Figure 5.16 will help to illustrate the life expectancy of a 50 Hz type fluorescent lamp in relation to the number of switchings in the course of a working day.

Free electrons emitted from the surface of the cathode proceed at high speed through the tube, producing a continuous flow of electrons which collide at random with atoms of mercury. Collision elevates the atoms of mercury to a temporarily higher energy state which is quickly lost and the atoms of mercury are returned to a normal stable state. In the passage of change from one state to another invisible *ultraviolet* light is emitted which excites the phosphorous coating on the inside of the tube generating *visible light*.

Different types of phosphorus emit various colours and a measured combination of phosphorus will produce the desired colour.

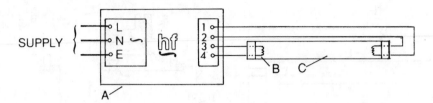

Figure 5.14. Fluorescent lighting control gear: high-frequency operation. (A, high-frequency ballast; B, heaters; C, lamp.)

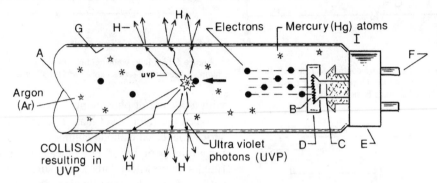

Figure 5.15. Inside a fluorescent tube. (A, glass tube; B, cathode heater; C, cathode lead wires; D, cathode shield; E, end cap; F, pins; G, coating of phosphor; H, visible light emitted by phosphor coating; I, glass support.)

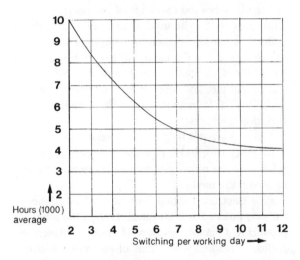

Figure 5.16. The life expectancy of a typical 50 Hz fluorescent lamp based on the number of switchings in a working day.

As current flowing through the tube has to be limited, control gear matching the length and wattage of the tube must be used. Makers of fluorescent lighting are able to offer tubes from 225 mm (9 imperial inches) to 2400 mm (8 imperial feet) in length and power outputs between 6 and 125 watts. The life expectancy of an average lamp,

other than a high-frequency type, is approximately 8000 hours based on a 3 hour switching cycle.

Ballast
A ballast unit is basically a wire wound inductance providing means of limiting current flow to the lamp without which the operational life of the lamp would be dramatically reduced.

It is essential to select the correct ballast to suit the lamp, taking into consideration both supply voltage and wattage. Control gear, when fitted or replaced, must not only be able to ignite the lamp but have the ability to limit the cathode-heating current. Unfortunately, energy is lost within a conventional ballast and this should be taken into account when energy calculations are made.

Incorrectly matched equipment will greatly shorten the life expectancy of the lamp or prevent the lamp from 'striking'. A major problem associated with wire wound inductance units is that of overheating. This will tend to reduce the life of both lamp and ballast due to current changes within the lamp. The operational temperature of a typical ballast is about 403 K (130 °C), but it will tolerate increases of temperature of up to 413 K, beyond which damage will be caused to both ballast and lamp.

Stroboscopic effects

A sinusoidal wave pattern (Figure 5.17) emitted from a single alternating light source can be detected by the human eye providing the frequency is not too high. This phenomenon is clearly demonstrated when low-pressure mercury vapour and other types of discharge lamps are connected to a 50 Hz AC supply. Incandescent (filament) lamps are also subjected to this problem but to a lesser degree as the light generated is a by-product of heat and the tungsten filament has insufficient time to cool to cause too much of a problem.

Physical effects

A varying intensity of light can produce both damaging and beneficial results as the following summary will show.

Damaging

1. Rotating machinery illuminated from a single light source will appear to have slowed down, changed direction of rotation or stopped. This is a potentially dangerous situation if the phenomenon is not understood or recognised.
2. Certain frequencies can induce degrees of drowsiness, headaches, eye fatigue and, in extreme cases, disorientation.

Beneficial

1. The stroboscopic effect can be harnessed to check and correct the speed of a CD or record player when the rotating element is equipped with special vertical strobe-lines

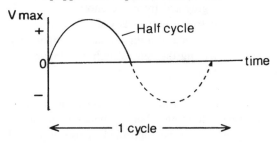

Figure 5.17. A 50 Hz sinusoidal wave pattern.

printed on the perimeter of the moving part.
2. A *stroboscope*, an instrument which enables rotating objects to be viewed, can be used to determine the speed of a rotating armature where, for example, it would be impractical to use a conventional rev-counter.

Counteracting stroboscopic effect

It is important to reduce or eliminate the stroboscopic effect for health and safety reasons especially, for example, in areas where rotating machinery is used.

In the following summary several ideas are advanced. Not all are practical for physical or economic reasons but they should be judged according to the requirements of the proposed installation.

1. The use of high-temperature filament lamps, especially when serving lathes, pillar drilling machines, etc. This will lessen the effect but will not eliminate it completely.
2. Direct current lighting. This will eliminate the effect but is not very practical.
3. The use of high-frequency fluorescent lighting. Special electronic control gear operates the lamp at a very high frequency of approximately 23 000 cycles per second (23 kHz). This considerably reduces the annoying ripple effect from 88 per cent experienced with standard control gear, to 20 per cent.

This type of lighting system is an ideal solution to the stroboscopic problem; however, high-frequency ballast units have radio interference suppression capacitors fitted between both phase, neutral and earth as Figure 5.18 will show. Unfortunately, 'switch-on-current' together with the minute but continuous current leakage to earth from the interference capacitor will trigger a residual current device, if incorporated within the circuit (Figure 5.19).

By adopting the following procedure the manufacturers advise how the problem of nuisance tripping may be overcome.
● The lighting arrangement should be divided over three phases using a 30

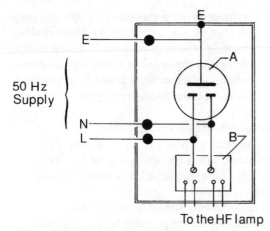

Figure 5.18. Control gear serving high-frequency fluorescent lighting. (A, radio interference suppression capacitor; L, phase conductor; N, neutral conductor; E, current protective conductor; B, electronic control gear.)

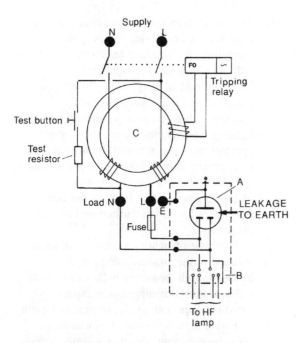

Figure 5.19. Control gear serving high-frequency fluorescent lighting will trigger a residual current device if incorporated in circuit. (A, radio interference suppression capacitor; B, electronic control gear; C, residual current device.)

milliamp surge-current-resistance, time-delayed residual current device (RCD)

● No more than 45 electronic ballast units per phase and an RCD connected, as illustrated in Figure 5.20.

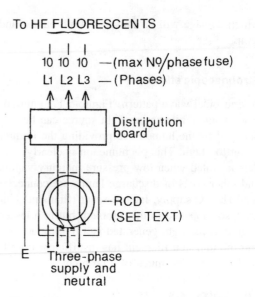

Figure 5.20. A way to overcome nuisance tripping.

Figure 5.21. A typical 400 volt warning sign.

4. The use of conventional fluorescent lighting divided over three phases.
    ● Each row of luminaires to be connected to a phase differing from an adjacent row.
    ● Switching arrangement either by a central contactor or by three separate banks of switches (one bank per phase), segregated from one another.
    ● Warning notices advising of the presence of a three-phase arrangement serving the general lighting (Figure 5.21).

Figure 5.22 illustrates in schematic form how a lighting scheme, divided over three phases, may be controlled by use of a contactor and a 5 amp lighting switch.

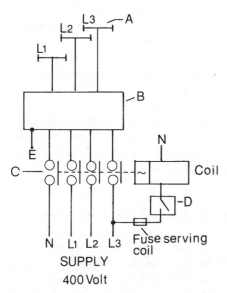

Figure 5.22. A contactor may be used to control a lighting arrangement originating from three separate phases. (A, lighting points; B, distribution centre; C, contactor; D, single-phase 5 amp control switch.)

High-pressure discharge lamps

A typical discharge lamp consists of a glass tube, known as an *arctube*, in which an electrode has been fitted at either end (Figure 5.23). Rare earth metals, such as *europium, lanthanum* or *lutetium* and mercury may also be added and the tube is pressurised with gases such as *argon* and *neon* before sealing. Lamps such as these will not normally start using a raw mains voltage but require an external ignitor to provide a high-voltage injection in order to set up an electromagnetic field. This excites ions present in the gas causing them to move at high speeds. Inevitably they collide with molecules of gas resulting in the emission of visible light, the colour of which will depend on the gas element or compound used.

Some systems are designed with an integral ignitor fitted within the body of the lamp. Lamps served with ignitors are readily identified by an internationally agreed capital Roman letter 'I' surrounded by a small equilateral triangle printed on the body of the lamp, as shown in Figure 5.24. Lamps designed to start using an external ignitor are annotated with the Roman capital letter 'E' within a small triangle, as illustrated in Figure 5.25.

Before full light output is achieved

Unlike incandescent and fluorescent lighting, the majority of discharge lamps are unable to obtain maximum illuminosity as soon as they are energised but require a short run-up period in order to achieve their full light output. During this period, which can last between 2 and 12 minutes depending on the type of lamp and control gear used, the initial starting current can rise to 190 per cent of the running current. It must be emphasised that not all categories of discharge lamps are designed to cater for such high starting currents, and on average 137 per cent can be expected.

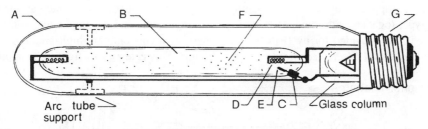

Figure 5.23. High-pressure discharge lamp. This type of lamp requires an external ignitor. (A, outer glass envelope; B, arc tube; C, resistor; D, electrode; E, auxiliary electrode; F, pressurised gasses; G, base cap.)

Figure 5.24. Lamps served with ignitors are readily identified by a capital letter 'I' surrounded by a small triangle.

Figure 5.25. Lamps requiring an external ignitor can be identified by a capital letter 'E' surrounded by a small triangle.

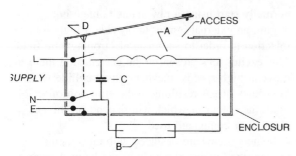

Figure 5.26. Wiring arrangement for a high-intensity mercury vapour discharge lamp. (A, ballast unit; B, lamp; C, power factor capacitor; D, switch-interlock.)

Older lamps

Towards the end of a lamp's working life, particle rectification can occur. This has the effect of producing a higher than normal running current and therefore it is wise to incorporate a suitable over-current device within the lamp ballast circuit to avoid possible damage to the control equipment.

High-pressure mercury vapour

Figure 5.26 illustrates a typical diagram designed for a mercury vapour lamp control circuit. This type of lighting is ideal for general use in factory and highway lighting schemes. Other forms are suited for many indoor applications. When compared to the domestic filament lamp, high-pressure mercury vapour lamps are found to be far more efficient, providing on average 47 lumens/ metre per watt. Reference is made to Table 5.1.

Figure 5.27 provides a see-through breakdown of the main integral components of this type of lamp. Lamps of this type, together with their associated control equipment, are provided for by British Standards or *International Electrotechnical*

Committee (IEC) specifications. Equivalent lamps and control gear manufactured by others are therefore interchangeable.

It is interesting to mention at this point that although the majority of mercury vapour lamps require external control gear to function, some types can be connected directly to the mains supply, thus avoiding the use of an external ballast unit. This type of lamp is equipped with a small integral filament which acts as a ballast and is ideally suited for domestic use.

Low-pressure sodium

Figure 5.28 highlights the integral wiring arrangement and control gear serving a typical low-pressure sodium vapour lamp. This type of lighting is ideal for heavy industrial applications such as steel mills and foundries or as a means to illuminate arterial roads and motorway junctions. Unfortunately the light emitted is monochromatic (one colour), under which colours are difficult to recognise. Compared with the domestic filament lamp they are extremely efficient, providing on average 145 lumens/metre per watt. Figure 5.29 provides a simple breakdown of the integral parts of a low-pressure sodium vapour discharge lamp. Associated control gear and lamps are all manufactured to British Standards and IEC Specifications and are completely interchangeable regardless of make.

High-pressure sodium

Figure 5.30 illustrates the internal wiring arrangements and accompanying control gear serving a conventional high-pressure sodium (HPS) lamp. This category of lighting is applicable for

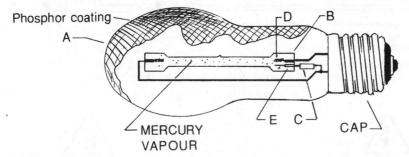

Figure 5.27. The basic components of a mercury vapour discharge lamp. (A, glass envelope; B, arc tube; C, resistor; D, electrode; E, auxiliary electrode.)

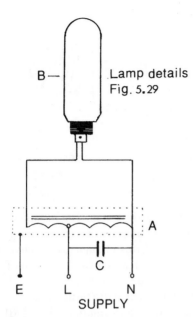

B— Lamp details
Fig. 5.29

A

C

E   L   N
SUPPLY

Figure 5.28. The internal wiring arrangements serving a low-pressure sodium vapour lamp. (A, leakage field transformer; B, lamp; C, power factor corrector capacitor; L. phase; N, neutral; E, current protective conductor.)

trade and industrial applications such as paper industries, power stations, cement works and pedestrian walkways. Other forms are ideally suited for horticulture and mining applications. Compared to the humble domestic filament lamp, which dispenses on average 7.6 lumens/metre per watt, the high-pressure sodium lamp is extremely efficient providing value for money by producing an average of 83.6 lumens/metre per watt. Figure 5.31 provides details of the internal structure of this type of lamp. As with other forms of discharge lighting, HPS

lamps are designed with two options; with or without internal ignitors. With this in mind care must always be taken to replace spent lamps with the correct type, details of which can been found on the glass envelope in the form of a triangular logo with the letter 'I' or 'E' ranged in the centre (Figures 5.24 and 5.25). HPS lamps are completely interchangeable with different makes and also conform to British Standards and IEC Specifications.

Neon lighting

High-voltage neon lighting is a specialised undertaking and therefore will only be dealt with in general terms. Figure 5.32 describes in schematic form how an installation may be wired to serve a

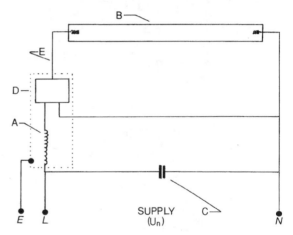

Figure 5.30. Internal wiring arrangements and control gear serving a high-pressure sodium vapour lamp. (A, ballast unit; B, lamp; C, power factor corrector capacitor; D, ignitor; E, short high-voltage ignitor lead.)

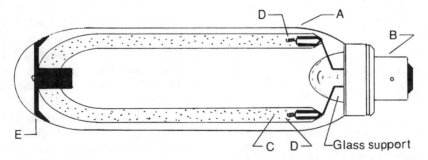

Figure 5.29. The integral parts serving a low-pressure sodium vapour lamp. (A, glass envelope; B, bayonet cap base; C, arc tube; D, electrode; E, internal arc tube support.)

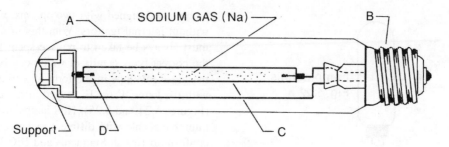

Figure 5.31. The integral parts serving a high-pressure sodium vapour lamp. (A, glass envelope; B, ES base; C, arc tube; D, electrode.)

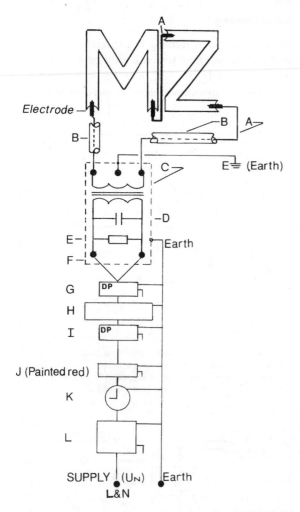

Figure 5.32. A small neon lighting installation. (A, high-voltage conductor; B, protective sleeve; C, high-voltage transformer; D, power factor correction capacitor; E, discharge resistor; F, steel housing; G, service isolator; H, distribution centre/over-current protection; I, lockable switch; J, fireman's switch; K, time switch; L, main switch.)

small neon advertising sign. The installation must be selected and built in strict accordance with the requirements of British Standards directive 559 (BS 559), copies of which can be ordered from most public lending libraries or obtained direct from the British Standards Institution. Reference is also made to *Wiring Regulation* 554–02–01.

Fireman's switch
Wiring Regulation 537–04–06 demands that an emergency fireman's switch be placed in circuit as a means of isolation in the event of fire. The following principles must be strictly adhered to:

1. The switch must be double pole when used with a single-phase supply and rated suitably.
2. The switch must be coloured red.
3. A sign measuring 150 mm × 150 mm must be placed on or near the switch, reading FIREMAN'S SWITCH. The size of the Roman capital letters should not be less than 1.25 cm (half an inch).
4. Both 'ON' and 'OFF' positions must be clearly marked; the off position being ranged at the top of the switch.
5. The switch must be equipped with a mechanical device preventing it from being switched on again after having being switched off.
6. The switch must be sited for the benefit of isolation in case of fire but out of reach of the general public. (In practice this would mean placing the switch at a height of 2.75 metres from the ground.)

Regulations governing low-voltage discharge lighting

Regulation 476–02–04 demands that self-contained luminaires, together with their circuit, be provided with appropriate means to isolate from the mains supply before gaining a working access to potentially live conductive parts. The following criteria are required to meet the conditions laid down by the *Wiring Regulations*.

1. A switched interlock serving the lid or cover which when opened will isolate the mains supply serving the luminaire before access is gained to any live or conductive part. This, of course, is in addition to the local lighting control switch.
2. A suitable and local means of isolating the circuit serving the luminaires which is in addition to the local light switch.
3. A lockable distribution centre or switch supplied with a dedicated key. Alternatively, a switch may be provided with a purpose-made removable handle which cannot be interchanged with others serving similar discharge lamp circuits.

Extra-low-voltage lighting

Tungsten halogen extra low voltage (ELV) is a very effective method of providing high-intensity illumination for display applications in commerce and may be used for decorative or general-purpose lighting in the home. When applied to recessed luminaires, lamps fitted have a clear toughened glass cover to protect against dust and humidity and general deterioration of the reflector. Figure 5.33 graphically depicts a typical extra-low-voltage tungsten halogen lamp.

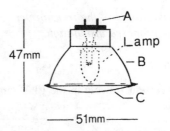

Figure 5.33. An extra-low-voltage tungsten halogen lamp. (A, contact pins; B, reflector; C, toughened glass cover.)

Methods of wiring

There are two methods of wiring, both of which should be considered.

1. Central transformer method
2. Multi-transformer method

Central transformer method

Figure 5.34 illustrates the basic wiring arrangements serving an extra-low-voltage (ELV) lighting system using the central transformer method. The following points should be borne in mind when carrying out such an installation.

(a) The volt-amp (VA) rating of the low-voltage transformer must match or over-accommodate the potential load of the secondary circuit, remembering that current flow in the secondary side of the transformer will be greatly increased due to the extra low voltage used.

   As an example, consider the following problem:

Six 12 volt, 50 watt tungsten halogen lamps are to be fitted in a shop window for display purposes. Calculate the current flowing in the ELV circuit and make recommendations

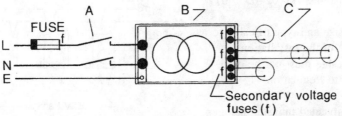

Figure 5.34. Basic wiring arrangements serving an extra-low-voltage lighting arrangement: central transformer method. (A, double pole switch; B, step-down transformer served with secondary voltage circuit fuses; C, luminaires.)

as to the rating of the transformer required for the proposed installation.

Solution:

Total current flowing in the ELV circuit

$$= \frac{\text{Total wattage}}{\text{Voltage}} \quad [5.1]$$

Applying known values:

$$\text{Total current} = \frac{6 \times 50}{12}$$

$$= 25 \text{ amps}$$

Calculating the size of transformer required:

Transformer rating = Secondary voltage

$$\times \text{ Secondary current}$$

$$= 12 \times 25$$

$$= 300 \text{ volt-amps (VA)}$$

$$[5.2]$$

A 400 VA transformer would be ideal. The additional rating could cater for future lighting requirements.

(b) A double pole switch should be wired into the low-voltage mains side of the installation. This will avoid the use of high-current switches in the ELV side of the circuit.

(c) Use standard low-voltage mains cable to serve the extra-low-voltage luminaires unless the circuitry can be completely segregated from other low-voltage mains circuits. Regulation 411–02–06 confirms.

(d) Do not connect exposed conductive parts serving a SELV system to earth; the low-voltage metal fitting is an example of this requirement. Regulations 411–02–05, –07 and –08 confirm.

(e) Select the correct size and type of cable to serve the ELV side of the installation. Metallic sheathed cable must not be used. Reference is made to Regulation 411–02–06(ii).

(f) Do not handle unprotected tungsten halogen lamps. Grease deposited from fingers will cause 'hot-spots' to develop on the glass envelope and reduce the life expectancy of the lamp. Lamps of this kind have a service life of between 2000 and 3000 hours.

(g) Select the correct voltage lamps to serve the installation. Generally ELV tungsten halogen lamps are available for 6, 12 and 50 volt supplies.

(h) The use of dimmer switches should be avoided as voltage variations will change the working temperature of the lamp and thus affect the halogen cycle, as discussed in previous paragraphs. If a dimmer switch has to be used and bulb blackening occurs as a result of the lamp burning for long periods at a reduced temperature, the lamp should be burnt using maximum working voltage for a short period of time. This will help to rectify the problem.

Multi-transformer method
Figure 5.35 illustrates the basic wiring arrangements serving an extra-low-voltage lighting system using the multi-transformer method.

One of the advantages of this system is that the ELV luminaires can be wired similarly to a low-voltage mains lighting installation, the only difference being that each lighting point is equipped with its own personal transformer. Another major advantage is that the size of cable used may be reduced to each transformer point. As the design voltage serving each transformer is considerably higher, current flow in the low-voltage mains circuit is proportionally lower.

Extra low voltage defined Extra low voltage is defined as a potential difference not exceeding 50

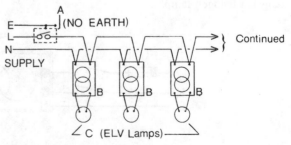

Figure 5.35. The multi-transformer method. (A, local switch; B, transformer; C, luminaires.)

volts AC or 120 volts DC when measured between both conductors or between the 'live' conductor and earth.

Low voltage defined Low voltage is defined as a potential difference exceeding extra low voltage but not exceeding 1000 volts AC or 1500 volts DC when measured between conductors or 600 volts AC, 900 volts DC when measurement is taken from the 'live' conductor and earth.

Safe handling and disposal of lamps

Handling

The following recommendations will provide information on the safe handling of lamps.

1. Read any instructions supplied with the lamp before fitting.
2. Do not handle tungsten halogen lamps with bare hands. This will avoid 'hot-spots' caused by grease-contaminated fingers.
3. Splashing water on lamps can create 'cool-spots' and cause the bulb to fracture.
4. Switch off the circuit locally before replacing a lamp. This is important especially when handling tungsten halogen high-temperature lamps.

5. Check that the lampholder is sound before attempting to fit a lamp and ensure that the lamp is securely placed in the lampholder. Tubular tungsten halogen lamps can sometimes be a problem to fit correctly.
6. Note that the design voltage of the lamp is suitable for the supply. An incandescent lamp with a working voltage of 110 volts can have the same bayonet cap base as a domestic 230–240 volt general-purpose lamp.

Disposal

Modern lamps contain many unpleasant compounds, elements and gases. Care must therefore be taken if breaking into small pieces is advised as a means of disposal. Most manufacturers recommend that defective lamps are disposed of strictly to national and *European Union* guidelines and regulations, copies of which may be obtained from local government offices.

To provide a modest degree of practical awareness of the hazards involved in the careless disposal of discarded lamps, Table 5.2 outlines in summary form a brief chemical profile of five familiar types of lamp, and methods of disposal.

Unsound or faulty lamps should be rejected but returned to the suppliers.

TABLE 5.2   The disposal of spent lamps: a chemical profile of five familiar types of lamp

Lamp	Ingredient(s)	Action/disposal
Incandescent	Glass, tungsten, argon and nitrogen gases; brass or aluminium	Do not break; place in a refuse container
General discharge	Glass, neon, mercury, argon and rare earth metals; alloys	Pressurised; not to be broken; adhere to national or local requirements
Sodium	Glass, neon, metallic sodium, metal components; alloys	Heat is produced when metallic sodium is placed in contact with moisture. Place up to ten discarded lamps into a large bucket. Break the lamps into small pieces. Gloves, eye protection and overalls must be worn. Fill the bucket half full with water using a hose pipe. Retreat to a safe place and after 5 minutes the active ingredients will become harmless
Tungsten halogen	Quartz glass, inert gas; bromine or iodine in small quantities; tungsten and insulating material	Pressurised bulb; do not break; adhere to disposal regulations
Fluorescent tubes	Very-low-pressure inert gases such as krypton or argon; glass, aluminium, insulating material, phosphor, liquid mercury and tungsten	Do not break; some tubes contain beryllium compounds which are poisonous; adhere to local or national disposal guidelines

Lamp bases

Each family of lamps has a distinctive base enabling instant recognition of the type to which they belong. Different bases also provide a degree of safety by assuring that lamps with special design characteristics are not mistakenly placed within the wrong holder. Figure 5.36 graphically shows the bases of some of the more familar types used in day-to-day electrical installation engineering work.

Application of lamps

Table 5.3 summarises in listed form a selection of non-specialised lamps used in electrical installation work. The life expectancy of a lamp is dependent on several factors such as switching cycles, location, temperature and the type of lamp used. Column 3 has been added as a guide based on data supported by the lamp manufacturer.

TABLE 5.3   Lamps and their application

Lamp	Typical application	Average lamp life in hours
Tungsten filament	General-purpose, domestic, decorative and commercial usages	1000
Striplight (filament)	Shaver lights, picture lights, etc.	1000
Tungsten halogen	General floodlighting, security lighting, traffic installations, building sites, decorative and domestic usages	1500–3000
Compact fluorescent	Domestic, commercial and industrial usages	Up to 8000
Standard fluorescent	Trade, industry, lecture rooms, schools, domestic kitchens, hospitals, surgeries, outdoor lighting, milking parlours, offices, public areas, etc.	Up to 8000
Discharge lighting	Building sites, trade and industry; floodlighting, traffic installations, public ares, street lighting, storage yards, etc.	8000
High-frequency fluorescent	Trade, industry, workshops and factories; not suitable where the installation is protected by a residual current device; see *Counteracting stroboscopic effect* on p. 113.	Up to 12 000

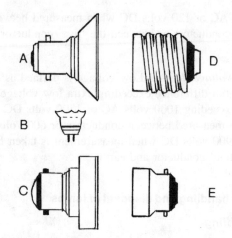

Figure 5.36. Lamp bases. (A, compact fluorescent; B, tungsten halogen; C, high-intensity discharge; D, high-wattage incandescent; E, general-purpose lamps.)

The lumen

The SI unit of luminous flux is the *lumen*, symbol lm, and can be defined as the quantity of light emitted from a body per second on a unit area placed at a unit distance from a source of light of 1 candela (symbol *cd*) intensity.

Lighting design engineers and lamp manufacturers apply luminous flux data to problems involving the design of lighting schemes and also during the drawing board stage of proposed new lamp developments. Some high-intensity discharge lamps have restricted burning positions and light emitted can sometimes vary if the lamp is not at the recommended angle.

Alternative switching arrangements

Contactor

Figure 5.37 describes in schematic form how a large working area, such as a warehouse or factory floor, served with low-pressure mercury vapour luminaires can be controlled by use of a simple industrial 5 amp lighting switch. Individual rows of luminaires can, if so required, be locally controlled once the general lighting has been switched on. An advantage gained by this method is that an operative is able to isolate all lighting by means of a single switch at the close of a working day.

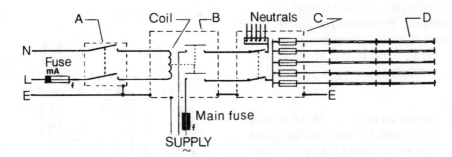

Figure 5.37. Large areas served by lighting may be controlled by means of a suitable 5 amp switch and contactor. (A, double pole switch; B, contactor; C, distribution centre; D, luminaires.)

Passive infrared (PIR) detector

When an automatic lighting control is requested, passive infrared switching can provide a solution. This type of automatic switching is ideal for security arrangements whether on-site or at home and will accommodate up to approximately 8–16 amps of resistive current.

Where security is of paramount importance, or where day-to-day running costs have to be taken into account, the passive infrared detector is a good choice for consideration. Figure 5.38 illustrates the basic wiring arrangements serving a typical PIR with optional over-ride switching facilities.

Time switch

Central heating controls are very dependent on reliable automatic switching arrangements and are generally to be found in the form of time switches. These are manufactured in a variety of shapes and sizes, some supplied with an integral clockwork mechanism which mechanically gains control upon mains failure. Figure 5.39 depicts an industrial time clock designed to automatically provide a switching mode for a heating system in a small factory.

Time switches can also be used to control outside security lighting arrangements or lighting schemes serving agricultural poultry houses.

Light-sensitive switch

Photo-electric switches are also known as *photo-cells* and there are many different types to be found both in trade and industry.

One of the more common applications is the control of individual street lights at dusk and dawn,

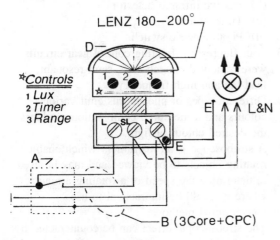

Figure 5.38. Lighting controlled by a passive infrared detector. (A, over-ride switch; B, cabling; C, floodlight; D, passive infrared detector.)

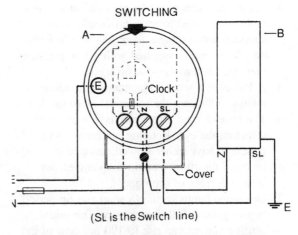

Figure 5.39. An industrial time clock. (A, time clock base; B, load. The switching action takes place between terminal L and SL.)

eliminating the need for expensive centrally controlled switchgear. Photo-cells can also be applied to control lighting displays in shop windows and may be found as part of a simple intruder alarm installation.

Construction

The device is constructed using a layer of *selenium* (a crystalline element which is electrically sensitive to light). The selenium is capped with a very thin see-through layer of the element *gold*. Light falling onto the cell produces a tiny voltage by the photovoltaic effect which activates the switching mechanism. Figure 5.40 shows the basic wiring arrangements required to provide automatic switching to an outside luminaire.

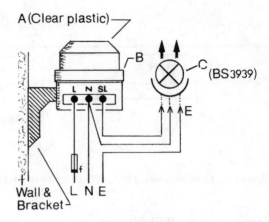

Figure 5.40. Control by means of a photo-cell. (A, clear plastic cover; B, photo-electric cell; C, luminaire. The switching action takes place between terminal L and SL.)

Summary

1. Lighting schemes for large projects are usually computer aided and are considered highly specialised.
2. Incandescent lamps are filled with a pressurised mixture of nitrogen and argon gases. Average life expectancy is 1000 hours.
3. Tungsten halogen lamps are not suitable for use with dimming circuits. A decrease in working voltage will cause the wall of the glass bulb to reduce significantly in temperature thus allowing 'blackening' to occur.
4. Fluorescent lamps can be mounted at any angle and are designed to work efficiently when supplied with voltages from 254 to 276 volts AC. The lamp is supplied with a low-pressure inert gas such as *krypton* or *argon*.
5. A ballast is a wound inductance providing means of limiting the current flow to the lamp.
6. The stroboscopic effect can have a physical effect on the eye. Rotating machinery will appear to have either slowed, stopped or reversed direction. Certain frequencies can induce degrees of drowsiness.
7. Discharge lamps require a short run-up period before gaining full illuminosity. The initial starting current can rise to 190 per cent of the running current.
8. There are two common methods of wiring

extra-low-voltage lighting installations:
   (a) Central transformer method
   (b) Multi-transformer method.
9. Splashing water on lamps can create 'cool-spots' and cause the bulb to fracture.
10. Each type of lamp has its own distinctive base enabling instant recognition.
11. The SI unit of luminous flux is the *lumen*, symbol lm.
12. The following alternative switching arrangements can be considered:
    (a) Central contactor
    (b) Passive infrared detector
    (c) Time switch
    (d) Photo-electric switch.
13. Four fluorescent lighting control gear circuits were reviewed: switch start, semi-resonant, lead-lag and high frequency.
14. Different types of phosphorus emit various colours and a measured combination produces the desired colour of light.
15. A stroboscope is an instrument which enables rotating objects to be viewed and may be used to determine the speed of a rotating armature where it would be impractical to use conventional methods.
16. The stroboscopic effect can be counteracted by use of high-frequency fluorescent lighting or by dividing a lighting arrangement over three phases.

17. Some high-pressure lamps are fitted with an internal ignitor and can be identified by a Roman capital letter 'I' placed within a small equal-sided triangle printed on the side of the lamp.
18. Some types of mercury vapour lamps can be connected directly to the mains supply. No external ballast is required.
19. Low-pressure sodium lamps are extremely efficient, providing on average 145 lumens/metre per watt.
20. A high-voltage neon lighting installation is a specialised undertaking and must be built to the requirements of BS 559.
21. Tungsten halogen lamps should not be handled with bare fingers. Hot spots can develop resulting from grease deposited on the glass envelope and reduce the service life of the lamp.
22. Modern lamps contain many unpleasant components. Spent lamps should be disposed of in accordance with published guidelines.
23. The average life expectancy of a fluorescent lamp is 8000 hours. This is based on a 3 hour switching cycle over a period of 24 hours.
24. Incandescent lamps were first produced in 1906. Carbon rod vacuum glow lamps were experimentally produced in the late 1850s and commercially developed in the 1880s.
25. Incorrectly matched control gear will greatly shorten the life expectancy of a fluorescent lamp or prevent it from 'striking'.

Review questions

1. List the mixture of two gases to be found in a modern incandescent lamp.
2. What is the practical reason for advising against using halogen lamps with dimming devices?
3. Other than high-frequency fluorescent lighting, what is the average service life of a standard fluorescent tube?
4. Confirm the following statements:
   (a) The use of a ballast unit is to provide a high voltage supply to allow the fluorescent tube to 'strike'. TRUE/FALSE
   (b) A typical discharge lamp consists of a glass tube known as an arc tube. TRUE/FALSE
   (c) Extra low voltage is defined as a potential difference not exceeding 50 volts AC or 120 volts DC. TRUE/FALSE
   (d) Direct current used for lighting arrangements will only partly eliminate the stroboscopic effect. TRUE/FALSE
5. What type of light is emitted from the inside of a fluorescent lamp?
6. When were halogen lamps first marketed in Europe?
7. List three automatic means of switching or controlling lighting arrangements.
8. List two practical manually operated methods of controlling a lighting arrangement.
9. Name the SI unit of luminous flux.
10. List three typical types of lamp bases.
11. British Standard 16 requires that a lamp should not deteriorate in illumination below a certain percentage after the first 750 hours. The percentage recommended is:
    (a) 75 per cent
    (b) 80 per cent
    (c) 85 per cent
    (d) 90 per cent.
12. Describe briefly the role of a ballast unit serving a low-pressure mercury vapour lamp.
13. High-pressure sodium lamps have a Roman capital letter 'E' or 'I' placed within a triangle printed on the glass envelope. Briefly state the reason.
14. How high should a fireman's switch be placed from the ground?
15. Why is it unwise to install a high-frequency fluorescent luminaire within an installation protected by a residual current device?

Handy hints

1. During the first fix stage of a domestic installation, place lighting cables serving wall-mounted luminaires at angle of approximately 70° to the ceiling. This will avoid accidental damage to the circuit through drilling when the wall has been plastered and electrical second fix work is being carried out.

Marking out an installation can be achieved by either measuring, templating or scribing.

2. Fluorescent slimline tubes fitted to rotary bi-pin base holders can be aligned with precision by positioning the small indent formed in the end-cap to face the floor.

3. A high-frequency fluorescent luminaire can trigger a residual current device into fault condition. This type of fitting has a radio interference suppressor connected between phase, neutral and earth.

4. Always provide written information and advice in simple non-technical terms to your customer when the installation is completed. This will help to develop goodwill.

5. The maximum life and efficacy of a high-intensity discharge lamp is dependent upon the supply voltage being ±5 volts of the rated voltage of the ballast unit.

6. It is advisable to install an upstairs domestic lighting circuit through an available service duct. This will avoid future nails and screws from damaging the cable once the installation has been handed over to the customer. It is easy to place fixings in the wrong place when there are no points of reference for guidance.

7. Whenever portable appliance testing is carried out, ensure that the flexible bonded cable serving the test equipment is securely attached to the metal frame of the appliance under test. Badly made connections will produce poor continuity values.

6 Basic electronics

In this chapter: Pioneers. Thermionic emission; triode valve. Semiconductors; crystal lattice; doping; semiconductor diodes. Testing. Resistors; colour code for resistors. Capacitors; types of capacitors. Electromagnetic induction. Graphical symbols used in electronics.

Pioneers

Towards the close of the nineteenth century in 1883, *Thomas Alva Edison* recognised that if an independent electrode was placed in a sealed incandescent lamp, current would flow between the hot filament and the unrelated electrode. For 22 years this remained a scientific curiosity until *Ambrose Fleming*, in 1905, designed and formulated the world's first workable rectifying valve. His achievement turned out to be a giant leap forward and paved the way for the commercial expansion of radio, television and radar.

Thermionic emission

When heated to approximately 2273 K (2000 °C) metals such as *thoriated tungsten* are able to release their surface electrons without difficulty. This phenomenon is known as *thermionic emission*; meaning the release of free surface electrons.

The components of a basic thermionic valve are enclosed in an evacuated glass envelope (air removed). The *cathode*, or negative electrode, made from nickel-coated tungsten, is designed to form a simple heating element. When heated, surface electrons escape from the cathode and are attracted to the positive collecting plate known as the *anode*. This often takes the shape of a metal cylinder adapted with cooling fins to help reduce surface temperature. The complete assemblage is known as

a *thermionic diode valve* whose symbol is schematically illustrated as Figure 6.1.

Space charge

The modern valve uses *oxide-coated tungsten* to form the cathode. Tungsten has a very high melting point, 3687.15 K (3414 °C), and the oxide coating allows good electron emission at temperatures between 1073 and 1273 Kelvin.

Figure 6.2 illustrates a simple directly heated diode valve where the heating element takes the form of the cathode. When connected to a suitably sized battery sufficient thermoelectric energy will be provided allowing electrons to be shed from the cathode and settle on the anode. Once saturated with negative electrons, the anode, being *negatively* charged, will oppose any further electron emission from the surface of the cathode. This is because like charges repel one another. (A similar phenomenon can be demonstrated with magnetic fields of the same polarity.) Excess electrons are then suspended in space between the two principal electrodes. This is known as the *space charge*. By increasing the voltage the suspended space charge

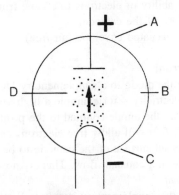

Figure 6.1. The thermionic diode valve. (A, anode; B, vacuum; C, cathode; D, electron flow.)

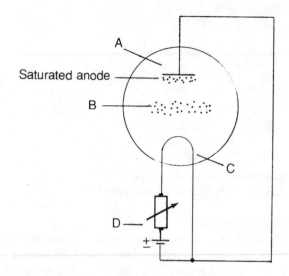

Figure 6.2. In this illustration the anode has been connected to the negative terminal. (A, negatively charged anode; B, space charge; C, cathode; D, battery and control.)

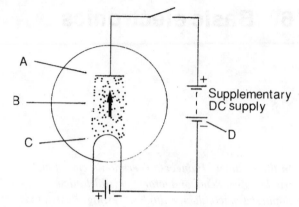

Figure 6.3. Free electrons from the cathode are drawn to the anode. (A, anode; B, electron flow; C, cathode; D, supplementary DC supply of electricity.)

will accelerate towards the anode only to be replaced by more electrons from the cathode once the anode is saturated.

It can therefore be said that *the diode current is directly proportional to the voltage applied*. It is interesting to note that a space charge will also form if the cathode is heated and there is no positive voltage serving the anode.

In practice, electron flow is governed by the following criteria:

1. Temperature.
2. Area of cathode.
3. Material composition of the cathode.
4. The ability of electrons to escape from the surface of the cathode.
5. An evacuated envelope (air free).

Anode current

Connecting the diode to a supplementary DC supply of electricity – for example a high-voltage battery – with the anode coupled to the positive pole (Figure 6.3) will allow free electrons emitted from the heated surface of the cathode to be drawn to the anode in a constant flow. This is known as *anode current*.

Isolating the supplementary supply will stop the flow of electrons and the anode will become negatively charged as previously described. This will then have the effect of repelling electrons emitted from the cathode, thus producing a space charge.

As the thermionic valve will only permit electrons to travel in one direction it may be used as a means of rectifying *alternating current*, AC, to *direct current*, DC.

Triode valve

In 1907 *Lee de Forest* added a third electrode to the basic diode. Taking the form of a spiral mesh or fine lattice, the structure was sandwiched between the anode and cathode. Valves such as these are known as *triodes*, illustrated in schematic form as Figure 6.4.

By making the grid completely negative with respect to the anode as described in Figure 6.5, electrons will amass as a space charge between the grid and cathode. This is known as the *cut-off voltage*. By studying Figure 6.5 it will be seen that electrons are repelled from both directions as like charges oppose each other. When the grid is made *less* negative in respect to the anode (Figure 6.6), a regulated electron flow from the cathode to the anode is permitted.

Application

Small variations of grid voltage can have a considerable effect on the anode current and should

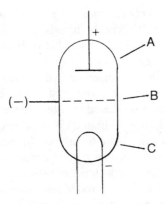

Figure 6.4. The triode valve. (A, anode; B, grid; C, cathode.)

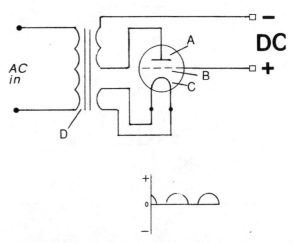

Figure 6.7. Simple half-wave rectification using a triode valve. (A, anode; B, grid; C, cathode; D, transformer.)

this type of valve be wired so that the output voltage from the anode of the first valve is connected to the grid of the *second* and so on, small currents can be amplified thousands of times.

The thermionic diode is able to handle far greater voltages than its semiconductor counterpart without breaking down and generally is able to operate at far higher temperatures. Diode valves are used for:

- amplification
- full-wave or half-wave rectification (Figure 6.7)
- generating high frequencies
- radar
- radio and television transmissions

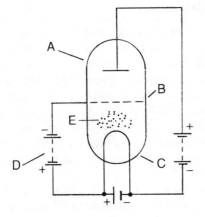

Figure 6.5. Electrons are repelled from both directions. This is known as the cut-off voltage. (A, anode; B, negative grid; C, cathode; D, DC power supply; E, suspended electrons.)

Indirectly heated valve

The modern thermionic valve is heated indirectly. The cathode is not an integral part of the heating element. Instead, an independent electrode in the shape of a metal cylinder is placed over the heating element and this is referred to as an *indirectly heated triode valve*. This is illustrated in schematic form as Figure 6.8.

Electrons emitted from the cathode are attracted to the anode across the evacuated (airless) envelope via the control grid. The grid is physically much closer to the cathode than the anode, therefore the grid's negative electrical potential has noticeably firmer command over electron flow than the anode

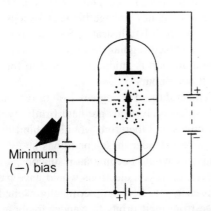

Figure 6.6. A regulated electron flow from the cathode to the anode is permitted when the grid is made less negative in respect to the anode. (A, anode; B, grid; C, cathode.)

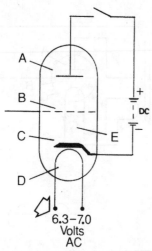

Figure 6.8. The indirectly heated valve. (A, anode; B, grid; C, cathode; D, heater; E, evacuated envelope.)

potential does. This allows the grid to be far more efficient in controlling valve current than the anode alone. This is known as the *amplification factor* of the valve, and is determined by physical and design features.

Semiconductors

Profile

It may be surprising to learn that the semiconductor has been in commercial production since the very early days of wireless transmission when it was applied to the pioneering 'cat's whisker' radio as a receiving diode.

Although not fully understood at the time, further research paved the way for the development of the transistor in 1948. Smaller and lighter than the thermionic valve, the semiconductor diode is not dependent on heat and can operate at far lower voltages. Diodes vary in size from one or two millimetres square when used for handling small amounts of current to sizes as large as oranges when used for power rectification.

Semiconductors are forged from the crystals of *germanium* (*Ge*) or *silicon* (*Si*). These are melted at the processing stage, purified and gas impregnated with an impurity.

At room temperature, pure semiconductor elements have a natural intrinsic electrical resistance lying between those of an insulator and a conductor and will *decrease* in resistance with a rise in temperature. They have what is described as a negative *coefficient of resistance*.

Atoms of both germanium and silicon are arranged so they form a cubic crystal lattice; each atom is dependent on the other. Both elements have four valence electrons within their respective outer shells, as shown in Figure 6.9(a) and (b), and are recognised as being neither a good conductor nor a bad insulator. Semiconductors are used in the development of solid-state diode transistors and complex integrated electronic circuitry.

Crystal lattice

Silicon is a very common element making up about 25 per cent of our planet's crust. Regrettably it never occurs in a pure and manageable state but is always combined with oxygen. The raw material from which semiconductors are made is gathered from quartz, sandstone and silicon-bearing rocks and is commercially prepared by energetic processes involving very high temperatures. Silicon has a melting point of 1690 K and a boiling point of 2870 K.

A semiconductor can be described as a crystalline element, structured in the form of a cubic lattice sharing pairs of valence electrons; one provided by each atom. This means that one valence electron from each atom is coupled with an electron from a bordering atom to form what is known as a *covalent bond*. This could be described as a common valency linkage between groups of atoms, as Figure 6.10 illustrates. Each atom is arranged at a fixed regular distance from its adjoining atom and is atomically bound to four of its nearest neighbours.

At a temperature of 0 degrees Kelvin (absolute zero), all valence electrons are held firmly by their covalent bonds and therefore would prohibit the flow of current. Under these conditions a semiconductor would become an insulator. At room temperature, however, electrons would have gained sufficient energy to become thermally excited and break free from their orbits to wander freely at random throughout the crystal lattice (Figure 6.11). Once an electron is freed, its parent atom will no

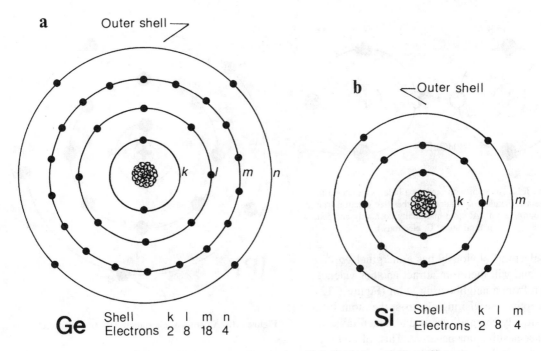

Figure 6.9. (a) An atom of germanium (Ge); (b) an atom of silicon (Si).

longer be neutral but will have a *net positive charge*.

Holes: missing electrons
The covalent gap that is developed by an electron leaving its parent atom is known as a *hole* and may

be said to be the place where an electron would be required in order to complete the crystal lattice. A balanced combination of holes and free electrons will provide an avenue for conductivity within a semiconductor. An incomplete atom without a full

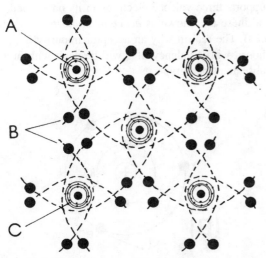

Figure 6.10. The covalent bond. (A, nucleus; B, valence electrons; C, inner electrons.)

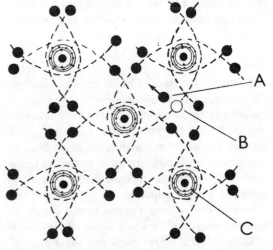

Figure 6.11. At room temperature valence electrons become thermally excited and break free from their orbits. (A, freed valence electrons; B, covalent gap; C, nucleus and inner electrons.)

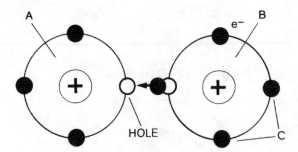

Figure 6.12. An atom withstanding a hole has a net positive charge and therefore will attract another free electron from a neighbouring atom. (A, atom marginally positively charged; B, neutral atom; C, electrons.)

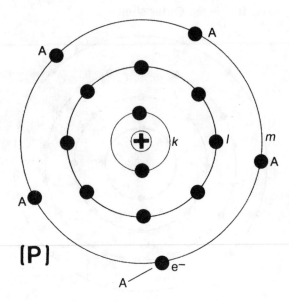

Figure 6.13. An atom of phosphorus (P). (A, valence electrons.)

set of electrons obviously has a marginal *positive charge* and will therefore attract another valence electron from a neighbouring atom (Figure 6.12). On accepting the visiting electron, the atom loses its 'positively charged' status and is neutralised; it is neither positive nor negative. This, of course, will leave a hole in the adjoining atom which will now have a clear positive charge; a charge sufficient to attract another electron from a neighbouring atom and the whole process continues *ad infinitum*.

It is fair to assume that internal current flow within a semiconductor results from a flow of negative electrons attracted to net positively charged atoms, and holes which appear to 'travel' in the opposite way. This is known as *intrinsic conduction*.

Doping

Impurities in the form of the elements *antimony* (Sb), *arsenic* (As) or *phosphorus* (P) are added to the semiconductor material during the production stage. All three elements have a valency of 5 (each has five free electrons), and therefore produce a surplus of one electron per atom in respect to the semiconductor material (Figure 6.13). This is known as a donor impurity.

Four of the five donor atoms form a covalent bond necessary for the construction of the crystal lattice. (This will be reviewed in a subsequent paragraph.) The fifth and excess electron is able to wander freely and is available for conduction.

A material impregnated in this fashion is known

as a *n-type semi conductor* and the process is known as *doping*. It is important that the impurity added must form part of the crystal lattice structure. It must occupy a place where an atom of silicon or germanium would normally have been, otherwise conductivity will be weakened.

Doping a semiconductor with either *gallium* (Ga), *boron* (B) or *indium* (In) produces what is called a *p-type material*. Each impurity element supports three valence electrons in its outer shell and these are known as *acceptor impurities* (Figure 6.14). The reason why an acceptor impurity is chosen with a valency of 3 will be dealt with in a subsequent paragraph.

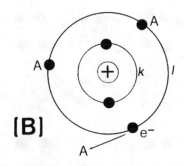

Figure 6.14. An atom of boron (B). (A, valence electrons.)

n- and *p*-type material

A semiconductor doped with either the donor element arsenic or antimony produces *n-type material*. Both elements have a valence of 5. This means that there are five electrons present in their respective outer shells. Figure 6.15 depicts the semiconductor germanium doped with the donor impurity antimony and highlights the excess electron.

When the semiconductor silicon has been contaminated with an acceptor impurity such as boron, an impurity element with a valence of 3, the resultant compound is known as *p-type material*. Figure 6.16 shows an atom of indium combined within the cubic lattice of germanium. It can be seen that the impurity atom has created holes within the lattice work. Remember that a hole is an area where an electron is required in order to complete the crystal lattice.

An incomplete atom surrenders its neutral status as it has a net positive charge and will attract an electron from a neighbour in order to stablise. This will cause a hole to appear within the valency shell of the adjoining atom and so the process continues.

Semiconductor diodes

The semiconductor diode may be compared to a non-return valve serving a domestic central heating system. It will allow water to flow through but not to return. The solid-state diode shares this fundamental principle. It will only allow the passage of current in one direction.

The simple diode can be heralded as the forerunner of many other sophisticated devices which are designed for use in modern electronic circuitry. Manufactured in various sizes, the diode's physical size echoes directly its current-carrying capacity. Small diodes are used in domestic radio and television receivers, whereas larger types are designed for use in power and rectification circuits.

The depletion layer

The anode is made from a shaped portion of p-type material and is factory fused to the cathode formed from an equally sized portion of n-type material. The two fused components form the diode.

The point and area at which both p- and n-type

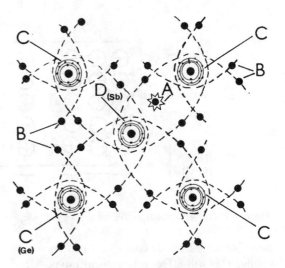

(Ge) Germanium atom .(Sb) Antimony atom

Figure 6.15. Germanium doped with the donor impurity antimony. This becomes n-type material. Only four of the these valence electrons will make the covalent bonds for the lattice. The fifth electron is able to wander freely throughout the crystal. (A, excess electron; B, valence electrons forming covalent bonds; C, an atom of germanium; D, an atom of the donor impurity antimony.)

semiconductors meet is called the *depletion layer* (Figure 6.17). This is so called because the area immediately adjoining the p–n junction has a shortage of electrons on the cathode side and a loss

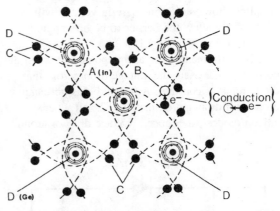

(Ge) Germanium atom. (In)Indium atom

Figure 6.16. Germanium doped with the acceptor impurity indium. This becomes p-type material. Only three of the four covalent bonds may be made. This results in holes being made providing an avenue for the conduction of electricity. (A, an atom of the acceptor impurity indium; B, hole; C, valency electrons forming covalent bonds; D, an atom of germanium.)

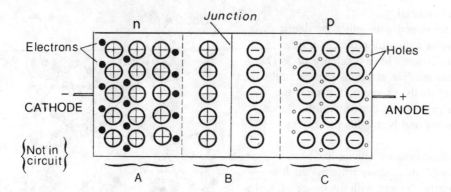

Figure 6.17. The depletion layer. (A, n-type material with free electrons; B, the depletion layer; C, p-type material with holes.)

of holes within the anode component of the coupling. This will effectively prevent n-type electrons moving across the junction to mix with p-type material. It is interesting to point out that the cubic lattice is unaltered throughout the union; only the impurity valence electrons vary on either side.

Applying a voltage: forward bias

Applying the minimum potential to a semiconductor diode will permit free electrons, or *carriers* as they are sometimes known, to accelerate from the battery plate towards the p–n junction. Some might gain sufficient energy to bridge the coupling; however, an increase in voltage will successfully allow the barrier to be broken and current to flow (Figure 6.18).

It is worth mentioning at this point that when the current is increased, the voltage drop across the diode remains constant. A forward bias germanium diode will drop approximately 0.2 volt whereas silicon diodes experience a voltage drop of around 0.6 volt. Semiconductors are not linear devices and therefore do not comply with Ohm's Law.

Reversed bias

Reversing the applied voltage (Figure 6.19), connecting the anode to the negative terminal of the battery and the cathode to the positive terminal, will attract electrons from the n-type material towards the positive battery plate. Holes from incomplete atoms in the p-type material are drawn to the negative plate as unlike charges attract. This in effect will widen the depletion layer until an electronic balance is reached and conduction is virtually stopped (Figure 6.20).

At this point the internal, resistance of a semiconductor diode can be extremely high. Should the voltage be increased even further, a total breakdown can be expected.

The zener diode

The zener diode, named after C.M. Zener (born 1905), is designed to operate in a reversed bias

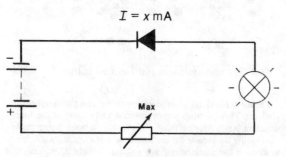

Figure 6.18. Forward biased circuit.

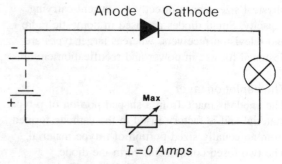

Figure 6.19. Reverse biased circuit.

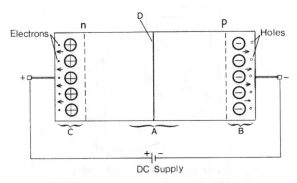

Figure 6.20. Widening the depletion layer will effectively stop the conduction of electricity. (A, extended depletion area; B, p-type material pulled away from the junction, D; C, n-type material pulled away from the junction, D.)

mode and will trigger at a predetermined potential. Voltage will be maintained even when the current flowing through the zenor diode is variable.

Figure 6.21 shows the basic requirements for a simple DC power source comprising:

- step-down transformer (A)
- four diodes connected in bridge formation (B)
- one electrolytic smoothing capacitor (C)
- one zener diode (D)
- one circuit fuse (F)
- double pole switch (G)

A zener diode is wired in circuit to regulate the voltage. When the bias is reversed it will stem current flow until the voltage applied reaches a critical programmed potential. Once reached, when typical values would be 5.6, 5.1 or 4.7 volts, the diode will permit current flow and maintain the selected potential even when the current flow through the diode is variable. Figure 6.22 illustrates a typical circuit incorporating a milliammeter.

Maintaining an arranged potential with a variable

current should not be taken to extremes. Like all components in electrical engineering, zener diodes have their practical working limitations and excess voltage will cause the diode to break down. Forward biasing will allow the zener diode to operate in the same way as any other solid-state diode.

Light-emitting diode (LED)

The light-emitting diode is used as a visual indicator and may also be moulded into display figures to serve calculators, alarm clocks and monitoring control equipment. Made from materials such as gallium arsenide (GaAs), the LED has proved to be far more dependable and durable when compared to the familiar incandescent lamp.

Wiring is carried out in forward bias and light is emitted at the p–n junction whenever holes and electrons recombine. The colour is dependent on the type of semiconductor used but, in practice, only red, yellow and green LEDs are available. The intensity of the luminescence is controlled by the bias current. Figure 6.23(a) and (b) provide a typical schematic arrangement showing a LED connected in series formation with a suitable resistor to maintain an operational current of

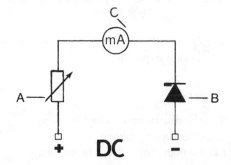

Figure 6.22. A zener diode wired in reversed bias mode. (A, variable resistor; B, zener diode; C, milliameter.)

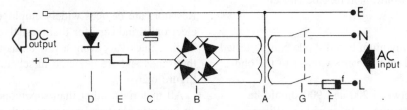

Figure 6.21. Full-wave rectification.

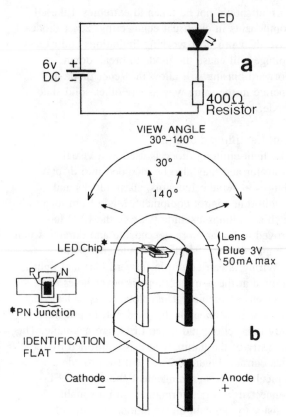

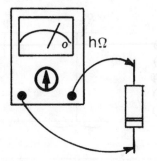

Figure 6.24. Testing forward biasing.

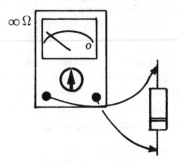

Figure 6.25. Testing reverse biasing.

Figure 6.23. An LED connected in series formation with a suitable resistor.

approximately 10 milliamps (0.01 A), calculated from the following expression.

$$R = \frac{U_s - 2}{I} \qquad [6.1]$$

where R is the resistance required in ohms
$\quad U_s$ is the supply voltage
$\quad I$ is the operating current in milliamps.

In practice it may not be possible to obtain the exact value of resistor required, and the nearest preferred value would therefore be chosen.

Testing

A silicon- or germanium-based diode can be tested using a standard multimeter switched to the ohms scale.

A value of between 200 and 900 should be expected when testing in forward bias (Figure 6.24), but when reviewed in reversed bias

(Figure 6.25), a value near to infinity can be predicted.

Although this test is very useful, it is unable to ascertain that the diode is accurately operating as devised. An extremely low meter reading when examined in both forward and reversed bias would indicate that the component had short circuited, whereas two infinitive values would suggest that the diode had developed an open circuit. In practice the faulty component should be carefully unsoldered and replaced using a diode matching the original in specification.

Resistors

Resistors are devices widely employed in the electronics industry for their resistive qualities. Their values vary from a few ohms to many thousands and are depicted in schematic form as Figure 6.26.

All materials, other than superconductors, have an intrinsic electrical resistance which will resist the flow of current.

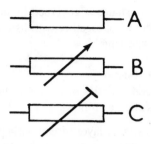

Figure 6.26. Resistors. (A, fixed; B, variable; C, preset adjustment.)

When a conductor's natural resistance is overcome, a small percentage of electrical energy is converted to heat or light or a combination of both. The electrical resistance of a material will depend on the following five considerations.

1. Length
2. Cross-sectional area
3. Temperature
4. Resistivity
5. Sometimes light (photo-cell).

Length
An increase in length will proportionally produce an increase in resistance.

Cross-sectional area
This is inversely proportional to length.

Temperature
Generally an increase in temperature will produce a proportional increase in resistance. Some materials such as carbon, germanium and silicon *decrease* in resistance with an increase in temperature. These are said to have a *negative coefficient* of resistance.

Resistivity
This is a constant; it represents the resistance offered by a cube of the material under review at a temperature of 273.16 K (0°C), and is expressed in ohms per meter. Resistivity is equal to the reciprocal of its conductivity. Hence,

$$\rho = \frac{R \times a}{l} \qquad [6.2]$$

where ρ (Greek lowercase 'rho') is the resistivity of the material under review
   R is the total resistance of the material
   a is the cross-sectional area of the material
and l is the length of the material under review.

Therefore, it can be said that the electrical resistance offered by a material is directly proportional to its length and inversely proportional to its cross-sectional area at a constant temperature.

In mathematical terms the resistance of a material (R) may be described as the ratio of the electrical potential difference (U) between the terminals of two conductors and the current (I) flowing through the conductors. Hence,

$$R = \frac{U}{I} \qquad [6.3]$$

As an example, consider the following:

A resistor of unknown value is connected across a potential difference of 100 volts and is found to be drawing a current of 0.25 amps. By use of Expression [6.3], calculate the value of the resistor.

$$R = \frac{U}{I}$$

$$= \frac{100}{0.25}$$

$$= 400 \text{ ohms}$$

Electrical resistance, quantified in ohms and named after *Georg Simon Ohm* (1787–1854), is mathematically symbolised by use of the Greek capital letter *omega* (Ω), and by the Roman letter R.

Types of resistors
Commercial resistors generally fall within the following categories:

1. Carbon resistor (Figure 6.27).
2. Wire-wound resistor (Figure 6.28).
3. Resistive film resistor (Figure 6.29).

Carbon resistor
The carbon-based resistor is constructed from granules of carbon mixed with an insulating

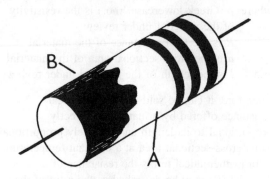

Figure 6.27. Carbon resistor. (A, thermoplastic coating; B, carbon granules plus filler and binder.)

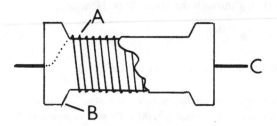

Figure 6.28. Wire-wound resistor. (A, resistance wire; B, ceramic former; C, connecting wires.)

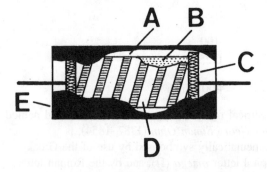

Figure 6.29. Resistive film resistor. (A, resistive film; B, ceramic rod; C, end caps; D, spiral grooves; E. protective coating.)

powdered filler and resin binder. The mixture is then fused together at high temperature; the value of the resistor is determined by the ratio of filler to carbon granules.

Wire-wound resistor

The wire-wound resistor is far more basic in construction and comprises resistance wire wound around a ceramic former and coated in a protective

material. The resistance is governed by the resistivity of the wire, together with its cross-sectional area and length.

Resistive film resistor

There are several varieties of resistive film resistors which are classified as follows:

1. *Carbon film* A layer of carbon is placed by thermal process onto a ceramic former.
2. *Metal film* Nichrome film is vacuum deposited onto a former.
3. *Metal glaze* A mixture of finely shredded metal and powdered glass is thermally fused to a ceramic former.
4. *Metal oxide* Minute particles of metal and oxide are combined together by heat treatment.

The declared resistance of any resistive film resistor is determined by the thickness of the film deposited, the ratio of conductive and non-conductive particles and the width of the spiral groove. During the processing stage a metallic film is first deposited and then a spiral groove is cut into the film to meet the required value of resistance. Table 6.1 profiles the principal characteristics and outlines practical applications for the three principal types of resistors found in industry today.

Power rating

Consideration must be given to the power rating required. Unfortunately the value is not usually indicated on the body of the component but the physical size of the resistor is a good indication of its potential power rating in watts. If in doubt, consult the manufacturer's product catalogue.

Alphanumerically coded resistors

These are coded with both letters and numbers. Resistors may either have their values printed on them conforming to the requirements of BS 1852, or may be colour coded by use of coloured bands wrapped around the body of the component. When the former is used, each value is annotated with two letters of the Roman alphabet (Figure 6.30). The first letter indicates the multiplier, taken from Table 6.2, whereas the second letter identifies the tolerance of the resistor. Although normally made

TABLE 6.1 The principal characteristics of resistors

Type	Comment/Application	Resistance range (ohms)	Working temperature (°C)	Temperature coefficient (parts per million/°C)	Power rating in watts (W)	Maximum voltage (V)
Carbon based	High frequency. Vintage radios	$10\,\Omega$ to $22\,\mathrm{M}\Omega$	−40 to +105	±1200	250 mW	150
Carbon film	Good characteristics. General purpose use	$10\,\Omega$ to $2\,\mathrm{M}\Omega$	−40 to +125	−750	250 mW	200
Wire wound	Electrical engineering. Used in large battery chargers	$0.25\,\Omega$ to $10\,\mathrm{k}\Omega$	−55 to +185	±200	2.5 W (min.)	415
Metal film	Expensive. Good stability compared to carbon base. Used for low power applications	$0.1\,\Omega$ to $100\,\mathrm{M}\Omega$	−55 to +155	±100	2 W	1.4 kV
Metal oxide	Poor stability compared to metal film	$10\,\Omega$ to $1\,\mathrm{M}\Omega$	−55 to +155	±250	500 mW	350
Metal glaze	Reliable but expensive	$10\,\Omega$ to $1\,\mathrm{M}\Omega$	−55 to +150	100	500 mW	250

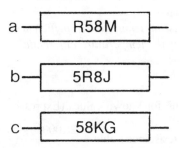

Figure 6.30. Resistors which are coded with both numbers and letters. (a) 0.58 ohms, 20 per cent tolerance; (b) 5.8 ohms, 5 per cent tolerance; (c) 58 000 ohms, 2 per cent tolerance.

TABLE 6.2 Resistor code; using both numbers and letters

First letter	Multiplier	Second letter	Tolerance level (±%)
R	1	F	1
K (kilo)	10^3 (×1000)	G	2
M (mega)	10^6 (×1000 000)	J	5
G (giga)	10^9 (×1000 000 000)	K	10
T (tera)	10^{12} (×1000 000 000 000)	M	20

to a nominal value a resistor will actually lie either side of this evaluation and it is this mid-point profile that is known as the *tolerance of the resistor*.

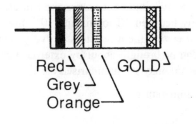

Figure 6.31. This resistor has a nominal value of 28 000 ohms at ±5 per cent tolerance level.

As an example: When measured by instrumentation a $100\,\Omega \pm 10$ per cent carbon-based resistor would be found to be between 90 and $110\,\Omega$.

Colour-coded resistors

A colour-coded resistor is designed to be read from left to right when viewed with the majority of coloured bands ranged to the left as shown in Figure 6.31. Each colour represents a number so that the value of the resistor may be identified.

- *Band 1* represents the first number.
- *Band 2* represents the second number.
- *Band 3* represents the multiplier or number of noughts following the first two numbers.

- *Band 4* (far right), represents the tolerance of the resistor.

As an example: By referring to Figure 6.31, Table 6.3 and Table 6.4 it will be seen that a resistor supporting a *red*, *grey*, *orange*, and gold band can expect to have a nominal value of 28 000 ohms at a ±5 per cent tolerance level.

A way to remember the colour code
In order to recall the sequence of colours used to identify the numerical value of a resistor; remember the following:

> Bleached Bread Rarely Offends Young Greedy Bluebottles Vandalising Grey Wheatmeal.

The capital letters represent the first letters of the colour code in numerical order.

Values
Resistors are manufactured in values ranging from 0.1 to 10×10^6 ohms but values used in electronics are usually between 10^2 ohms and 10^5 ohms. In a controlled environment resistors are very stable but change can be brought about by external factors. These factors may be summarised as:

1. Temperature
2. Age
3. Humidity

Temperature
Both thermal expansion and the temperature coefficient of a resistor must be considered. Quoted in *parts per million* for each degree Celsius and written as *ppm/°C*, the temperature coefficient for several varieties of commercial resistors may be found in Table 6.1.
   As an example, consider the following problem:

A 28 000 ohm carbon film resistor has a temperature coefficient of −750 ppm/°C and is subjected to an increase in temperature of 15 °C. Calculate the changed value of the resistor by using the following expression:

$$\text{Changed resistance value} = \frac{R \times \alpha \times {}^\circ t}{10^6} \quad [6.4]$$

TABLE 6.3   Resistor and capacitor colour coding

Colour code: bands 1 to 3	Number	Multiplier (band 3)
Black	0	1
Brown	1	10
Red	2	100
Orange	3	1000
Yellow	4	10 000
Green	5	100 000
Blue	6	1000 000
Violet	7	—
Grey	8	—
White	9	—

TABLE 6.4   Resistor colour coding: percentage tolerance

Coloured band 4 only	Tolerance level (±%)
Brown	1
Red	2
Gold	5
Silver	10
No colour	20

where R is the value in ohms of the resistor under review
   α *is the temperature coefficient in ppm/°C*
   °t is the temperature differential.

Solution:
Substituting for known values (Expression [6.4]):

$$\text{Changed resistance value} = \frac{28\,000 \times (-750) \times 15}{10^6}$$

$$= -315 \text{ ohms}$$

It should be remembered that carbon has a negative coefficient of resistance.

New value = Original value ± Changed value

$$= 28\,000 - 315$$

$$= 27\,685 \text{ ohms}$$

Age
Due to physical and chemical changes occurring within the body of a resistor, variations in nominal value can be expected with age. Old or second-hand resistors should be tested and replaced if necessary.

Humidity

A resistor will change in value if contaminated with water vapour. It is therefore advisable to keep stock items from potential sources of liquid pollution, humidity or dampness.

Resistive malfunction

Resistors, as with any other electrical component, will break down from time to time and either will be left in open circuit mode, causing disturbance to audio circuits, or will increase in resistance value, so high as to cause operating problems to the equipment they serve.

High value failure can be attributed to:
1. Movement of carbon granules.
2. Extraneous heat.
3. Moisture absorption causing the resistor to become bloated.
4. Excess voltage.

Open circuit failure can be caused by:
1. Mechanical stress.
2. Broken connections.
3. Excess heat leading to the interior of the component being burnt out.
4. Soldered joints failing under excess heat.
5. Excess load.
6. Broken wire in a wire-wound resistor.
7. Bad manufacture.
8. Disintegration of metal film.
9. Electrolytic action after moisture has been absorbed.
10. Impurities causing crystallisation within a wire-wound resistor.
11. Mechanical failure.

Resistors connected in series formation

Resistors connected one after the other, enabling current to flow through each component in turn, are said to be connected in *series formation*. The total resistance of the formation may be calculated by *adding* the sum of the individual values. Hence:

$$R_t = R_1 + R_2 + R_3 + \dots \qquad [6.5]$$

where R_t is the total resistance in ohms
and R_1, R_2, R_3 are the values of the individual resistors.

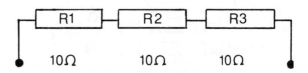

Figure 6.32. Resistors connected in series formation.

As an example, consider the following:

Calculate the total resistance of three 10 ohm resistors connected in series formation as illustrated in Figure 6.32.

Solution:
Referring to Expression [6.5]:

$$R_t = R_1 + R_2 + R_3$$

Substituting figures,

$$R_t = 10 + 10 + 10$$

$$= 30 \text{ ohms}$$

Resistors connected in parallel formation

Resistors connected so that each individual component receives exactly the same value of voltage are said to be connected in *parallel formation*. The total resistance when connected in this fashion may be calculated by use of the following expression:

$$\frac{1}{R_t} = \frac{1}{R_1} + \frac{1}{R_2} + \frac{1}{R_3} + \dots \qquad [6.6]$$

where R_t is the total resistance in ohms
and R_1, R_2, R_3 are the values of the individual resistors.

As an example, consider the following problem:

Find the total resistance of three 10 ohm resistors connected in parallel formation as illustrated in Figure 6.33.

Solution:
Referring to Expression [6.6], and substituting for figures:

$$\frac{1}{R_t} = \frac{1}{10} + \frac{1}{10} + \frac{1}{10}$$

First find the lowest common multiple; which is 10,

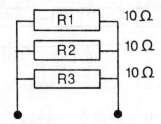

Figure 6.33. Resistors connected in parallel formation.

$$\frac{1}{R_t} = \frac{1+1+1}{10}$$

$$= \frac{3}{10}$$

Cross-multiplying:

$$3 \times R_t = 1 \times 10$$

Dividing both sides of the equation by 3:

$$R_t = \frac{10}{3}$$

$$= 3.333 \text{ ohms}$$

Product over the sum method

A quicker means to calculate *two* known resistors placed in parallel formation is by use of the *product over the sum method*. Hence,

$$R_t = \frac{R_1 \times R_2}{R_1 + R_2} \qquad [6.7]$$

This expression is only suitable for *two* resistors connected in parallel formation. When a problem involves more than two components, Expression [6.6] must be used.

Resistors placed in parallel formation will produce a far different resolution than from connecting the same components in series. The sum of any parallel configuration will always be less than the smallest valued component. Hence, when two large and one small valued resistor are wired in this manner the resultant sum will always be mathematically subordinate to the smallest value.

Resistors connected in series–parallel formation

When problems arise involving resistors connected in both series and parallel formation it is sensible to

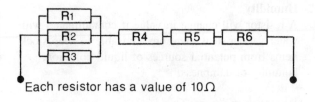

Each resistor has a value of 10Ω

Figure 6.34. A series–parallel formation of resistors.

resolve the parallel section first. Once the total resistance has been reduced to a single value the resultant sum can be added to the series combination of the problem. The values of each individual resistor simply need to be added together to evaluate the total resistance of the combination.

As an example, consider the following:

Find the total resistance in ohms of a series–parallel formation comprising three 10 ohm resistors connected in parallel together with three 10 ohm resistors placed in series formation, as illustrated in Figure 6.34.

Solution:

Referring to Expression [6.6] and the parallel element, and substituting figures:

$$\frac{1}{R_t} = \frac{1}{10} + \frac{1}{10} + \frac{1}{10} \quad (\textit{Lowest common multiple is 10})$$

$$= \frac{1+1+1}{10}$$

$$= \frac{3}{10}$$

Cross-multiply,

$$3 \times R_t = 10$$

Divide both sides of the equation by 3 to present the equation in terms of R_t:

$$R_t = \frac{10}{3}$$

$$= 3.333 \text{ ohms}$$

The total resistance of the parallel formation, 3.333 ohms, may now be added to the series leg of the problem as shown in Figure 6.35.

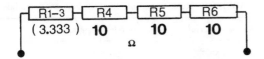

Figure 6.35. The total resistance of the parallel formation may be added to the series section of the problem.

Grand total = Sum of the parallel leg +

Sum of the series leg

$$= 3.333 + 10 + 10 + 10$$

$$= 33.333 \text{ ohms}$$

Capacitors

A capacitor is an electronic device which will enable an electric charge to be stored when both its plates are at opposite potentials. This property is known as *capacitance*.

A crude capacitor may be made from an insulating compound, such as mica, forming a dielectric buffer sandwiched between two matching metal plates (Figure 6.36). When a direct current potential is applied to the plates, current will appear to flow but will quickly subside once the capacitor is fully charged. By removing the source of electrical energy and linking the plates by means of an insulated test resistor, current will once again flow but this time in the reverse direction. This is often accompanied with a modest illuminated display, immediately followed by an audible report depending on the value of the capacitor and the test resistor.

Unit

The unit of capacitance is the *farad*, whose symbol is F. This is an extremely large unit to use and so for practical purposes it is subdivided into smaller parts which mathematically are easier to handle. A *microfarad* is one-millionth of a farad and is recognised by the symbol µF. The *picofarad*, also found in electronic circuitry, is equivalent to one-millionth of a microfarad, and is recognised by the symbol pF.

Named after *Michael Faraday* (1791–1867), the farad is defined as the capacitance developed between two plates charged with 1 coulomb of electrical energy.

The *coulomb*, symbol Q, and named after *Charles Augustin Coulomb* (1736–1806), is determined as a measure of negative electrons transposed by a current of 1 amp in 1 second of time. This has been calculated as 6.24×10^{18} (or 624 followed by 16 noughts) electrons passing a given point in one second of time. Hence

$$Q = I \times t$$

where Q is the quantity in coulombs
I is the current in amps
t is the time in seconds

Relative permittivity

Figure 6.37 depicts a rudimentary air-type capacitor connected to a DC source of electricity. Under these conditions the capacitance developed would

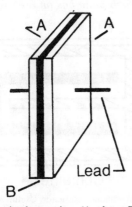

Figure 6.36. A simple capacitor. (A, plates; B, dielectric.)

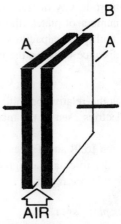

Figure 6.37. A simple air-type capacitor. (A, plates; B, air dielectric.)

be very modest. However, if a tailored sheet of mica was to be sandwiched snugly between the two plates it would be found that the value of the capacitance would be considerably increased.

The ratio of the capacitance of a given capacitor with a specific dielectric and the capacitance of the *same* capacitor with only air between the plates is known as the *relative permittivity* of the material under review. This is represented by the lowercase Greek letter *epsilon*, ϵ. Hence,

$$\epsilon_r = \frac{\epsilon}{\epsilon_0} \qquad [6.8]$$

By cross-multiplying to bring the expression in terms of ϵ:

$$\epsilon = \epsilon_r \, \epsilon_0 \qquad [6.9]$$

where ϵ_r is the relative permittivity
ϵ_0 is the permittivity of free space ($8.854 \times 10^{-12}\,\mathrm{F\,m^{-1}}$)
ϵ is the permittivity.

Capacitance developed

The capacitance developed by a given capacitor can be evaluated by use of the following expression:

$$C = \frac{A \times \epsilon_r \times \epsilon_0 \times (n-1)}{d} \qquad [6.10]$$

where C is the capacitance in farads
A is the area of one side of each plate in $\mathrm{m^2}$.
ϵ_r is the relative permittivity
ϵ_0 is the permittivity of free space
n is the number of plates (the term $n-1$ is used for multiplate capacitors)
d is the distance between the plates in metres.

Table 6.5 defines the relative permittivity of a selection of dielectrics used in the manufacture of capacitors.

From Expression [6.9] and Tables 6.5 and 6.6 it is clear that in order to obtain a high capacitance, the following criteria are required:

1. *Generous plate area.*
2. *Minimum distance between the plates.*
3. *High dielectric constant, ϵ.*

TABLE 6.5 Relative permittivities of dielectric materials

Dielectric material	Relative permittivity
Tantalum oxide	27.0
Ceramic	7.0
Porcelain	6.0 to 8.0
Bakalite	4.5 to 7.5
Waxed paper	4.0
Shellac	3.0 to 3.7
Mica	3.0 to 7.0
Rubber	2.8 to 3.0
Polypropylene	2.5
Paraffin wax	2.0 to 2.5
Air	1.000 59
Vacuum	1.0

4. *Suitable dielectric insulating properties to meet the working voltage while maintaining a minimum distance between the plates (see Table 6.6).*

Figure 6.38 schematically outlines a compound capacitor constructed with five plates. By referring to the illustration it will be seen that each side of *Plate P2* is in close contact with the dielectric, whereas only four out of a possible six sides of *Plate P1* are in partnership with the dielectric. As the outer plate is only served with a dielectric on one side, the total working surface area must therefore be one plate less than the number of declared plates. Hence,

$$A \times (n-1) = \text{Useful working plate area} \qquad [6.11]$$

As an example, consider the following and Figure 6.38.

A multiplate capacitor is designed with five plates. One side of each plate has an area of 600 mm²

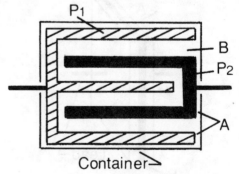

Figure 6.38. A compound capacitor. Useful area = $A(n-1)$. (A, plate area; B, dielectric.)

(0.06 m²) and is separated by a dielectric of waxed paper 0.0002 metre thick. The relative permittivity of the waxed paper is given as 4. Calculate the total capacitance generated in farads.

Solution:
Refering to Expression [6.10]:

$$C = \frac{A\,\epsilon_r\,\epsilon_0\,(n-1)}{d}$$

Known values:

A = Area, 0.06 m
ϵ_r = Relative permittivity of 4
ϵ_0 = Permittivity of free space, 8.854×10^{-12}
d = Distance between the plates, 0.0002 m
n = Number of declared plates, 5

Substituting figures:

$$C = \frac{0.06 \times 4 \times (8.854 \times 10^{-12}) \times 4}{0.0002}$$

$$= 4.249\,92 \times 10^{-8}\ \text{F}$$

or

$$C = 0.000\,000\,042\,9\ \text{farad}$$

Capacitance explained

Figure 6.39 schematically depicts a simple capacitor connected to the terminals of a low-voltage battery.

Basic principles support the concept that electron flow is a stream of negatively charged subatomic particles. As long as an electrical pressure is applied to the capacitor the positively connected plate will continue to attract electrons from the negatively connected plate and allow current to flow. The process will continue until the voltage stored across the capacitor is equal to that of the battery.

This may be expressed more clearly in summary and by reference to Figure 6.40.

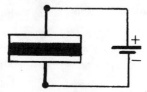

Figure 6.39. A simple capacitor connected to a low-voltage battery.

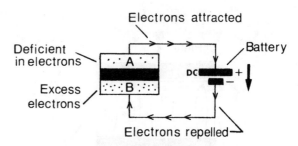

Figure 6.40. Capacitance explained.

1. Electrons from capacitor, *Plate A*, are attracted to the positive plate of the battery.
2. Once within the battery, electrons are thrust by electromotive force to the negative plate of the battery.
3. On arrival, electrons are then swiftly repelled towards *Plate B* of the capacitor as like charges repel one another.
4. Electrons which were once within *Plate A* are now within *Plate B*. This produces an excess of negatively charged electrons in capacitor *Plate B* and forms an electrostatic field which is equal in force to the supply voltage.
5. Current will cease to flow once a potential difference, equal to the voltage of the battery, exists across the plates.

Dielectric stress

As a potential difference exists between both plates of the capacitor an electrostatic field is produced which is equal to that of the supply voltage. Disconnecting will leave the capacitor in a fully charged condition. At this stage, electrons forming the dielectric are repelled from the side nearest to the negative plate and attracted towards the edge of the dielectric adjacent to the positive plate. The electrical pressure between the two opposing plates produces an elliptical deformity within the orbits of the electrons forming the sandwiched dielectric material, as shown in Figure 6.41.

This atomic deformity is known as *dielectric stress* and is directly proportional to the applied voltage. Increasing the voltage beyond the strength of the dielectric will cause a physical breakdown to occur, inevitably resulting in a short circuit.

Table 6.6 tabulates a short list of dielectric

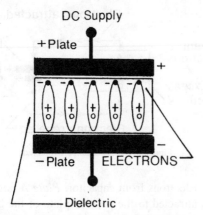

Figure 6.41. The electrical pressure between the two plates produces an elliptical deformity within the oribits of the electrons forming sandwiched dielectric material.

TABLE 6.6 Dielectric strengths

Dielectric material	kV/mm thickness	Application
Mica	40–200	High frequencies
Waxed paper	40–60	Low frequencies
Glass	5–30	High frequencies
Paper	4–10	Low frequencies
Air	3–6	Radio tuning

strengths of material used in the production of capacitors. Values are in kilovolt per millimetre thickness.

Types of capacitors

There are many different types of capacitors in use today. Some are used for electronic work, while others are utilised in electrical engineering.

Listed in Table 6.7 are six of the more familiar varieties of capacitors found throughout industry.

Figure 6.42 illustrates a typical paper-insulated capacitor widely used in starting circuits for single-phase motors and as a means of power factor improvement.

Alternative layers of aluminium foil and a matching dielectric are rolled to form a spiral cylinder. Leads are bonded to the aluminium capacitor plates, and the complete assemblage is placed in a suitable plastic or metal container for protection.

Capacitor colour coding

Regrettably, capacitor colour coding can vary slightly, depending on the source of manufacturer but generally attempts are made to follow the same criteria as laid down for resistors (Table 6.3). Large industrial cylindrical-type capacitors have their values printed on the side of their protective container. It is only the smaller types which need to be colour coded.

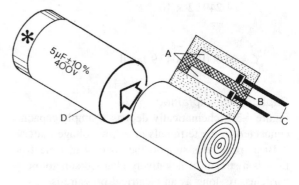

Figure 6.42. A paper-insulated capacitor. (A, plates; B, dielectric; C, leads; D, container.)

TABLE 6.7 Standard capacitors
(pF = 10^{-12} F, nF = 10^{-9} F, μF = 10^{-6} F. V = volts, kV = volts × 10^3)

Type	Application	Voltage range DC	Voltage range AC	Tolerance (%)	Range in value
Ceramic	Temperature compensating	60 V to 10 kV	250 V	±20	5 pF to 1 μF
Silver mica	Tuned circuits. Filters	60–600 V	—	±0.5	5 pF to 10 nF
Paper (foil)	Motor starting. Power factor correction	600 V	250 to 500 V	±10	10 nF to 10 μF
Etched electrolytic	Smoothing circuits	6–100 V	Polarised	+50	1 nF to 10^6 μF
Polystyrene	Filters. Tuned circuits	50–500 V	150 V	±1	50 pF to 0.5 μF
Polypropylene	Mains suppression. Motor start and run.	1.25 kV	600 V	±5	1 nF to 100 μF

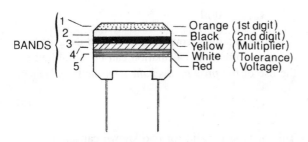

Figure 6.43. Capacitor colour coding.

As a general guide, the following arrangements apply to colour-coded capacitors. Reference is made to Figure 6.43, a typical low-valued non-polarised capacitor used in electronic circuitry, value-coded by use of coloured bands.

1st Band: This is positioned furthest from the point of connection and represents the first digit of the value of the capacitor.
2nd Band: Second digit of the capacitance value.
3rd Band: The multiplier or the number of 'zeros' following the first two digits.
4th Band: Percentage tolerance (see Table 6.8).
5th Band: Operating voltage (red up to 250 volts; yellow, 400 volts).

As an aid to clearer understanding, Figure 6.43 may be evaluated with help from Tables 6.3, 6.7 and 6.8.

1st Band: Orange. First digit from Table 6.3 is 3
2nd Band: Black. Second digit from Table 6.3 is 0
3rd Band: Yellow. The multiplier, taken from Table 6.3, is 10 000

TABLE 6.8   Capacitor tolerance colour code

Colour code	Percentage tolerance ($\pm$)
Brown	1
Red	2
Orange	2.5
Green	5
White	10
Black	20

4th Band: White. The percentage tolerance taken from Table 6.8 is 10 per cent
5th Band: Red. The working voltage; given as up to 250 volts.

By arranging the values obtained in coloured sequence and noting the characteristics of the capacitor (Table 6.7), it will be found that the capacitor has a value of 30 000 pF, 10 per cent tolerance with an operational voltage of up to 250 volts. To express in microfarads, divide the value of capacitance obtained by 1 000 000. Hence, 0.03 μF.

Capacitor failure
From time to time capacitor malfunction becomes inevitable. Even thoroughly planned maintenance programmes sometimes fail to detect potential problems. Listed as Table 6.9 are three of the more familiar and commonplace capacitors, profiling causes of failure.

Capacitors in series formation
When capacitors are connected in series formation (Figure 6.44), the total capacitance is reduced in value. This is *opposite* when compared to resistors

TABLE 6.9   Capacitor malfunction

Capacitor type	Problem	Suggested cause
Mica	1. Plates short circuiting	1. Movement of silver due to moist or humid conditions
	2. Open circuit or intermediate open circuit	2. Mechanical damage
Ceramic	1. Short circuit	1. Dielectric ruptured
	2. Inconsistent value	2. Fault condition between the dielectric and plates
	3. Open circuit	3. Broken connection
Paper foil, oil impregnated	1. Short circuit	1. Mechanical shock. High ambient temperature causing high pressure within the capacitor
	2. Intermediate open circuit	2. Loose electrical connection, inside or out. Mechanical damage
	3. Bloated or distended	3. Very high temperature. Breakdown of insulation

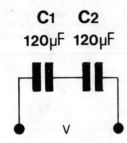

Figure 6.44. Capacitors connected in series formation.

coupled in a similar fashion and also a convenient way to remember.

Given that C_t is the total capacitance of a given circuit or configuration; C_1 is the value of the first individual capacitor and C_2 is the value of the second individual capacitor, then

$$\frac{1}{C_t} = \frac{1}{C_1} + \frac{1}{C_2}$$ 　　　　[6.12]

As an example, consider the following and Figure 6.44:

Evaluate the total capacitance of two 120 μF capacitors connected in series formation.

Determine the safe working voltage if each capacitor is rated to operate at 50 volts.

Solution:
Referring to Expression [6.12] and substituting values,

$$\frac{1}{C_t} = \frac{1}{120} + \frac{1}{120}$$

Determine the lowest common multiple (=120).

$$\frac{1}{C_t} = \frac{1+1}{120}$$

$$= \frac{2}{120}$$

Cross-multiply to bring the equation in terms of C_t:

$$2 \times C_t = 120$$

Divide each side of the equation by 2:

$$C_t = \frac{120}{2}$$

Total capacitance, therefore:

$$C_t = 110 \ \mu F$$

When evaluating a combination of two capacitors in series formation, calculation can be simplified by the use of the *product over the sum* method. Hence,

$$C_t = \frac{C_1 \times C_2}{C_1 + C_2}$$ 　　　　[6.13]

This method is valid for two known capacitors connected in series formation. Should there be more than two, Expression [6.12] must be used.

Safe working voltage
Each capacitor has a maximum working voltage of 50 volts but when connected in series formation the working voltage is doubled to 100 volts.

Capacitors in parallel formation
Placing capacitors in parallel formation (Figure 6.45) will effectively enlarge the area of plate and therefore increase the value of the total capacitance.

Given that C_t is the total capacitance, C_1 is the value of the first individual capacitor and C_2 is the value of the second individual capacitor, then

$$C_t = C_1 + C_2$$ 　　　　[6.14]

As an example, consider the following and Figure 6.45:

Evaluate the total capacitance of two 120 μF capacitors wired together in parallel formation.

Determine the safe working voltage when capacitor C_1 is rated at 50 volts while capacitor C_2 is valued at 10 working volts.

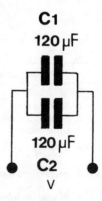

Figure 6.45. Capacitors connected in parallel formation.

Solution:

Referring to Expression [6.14] and substituting known values,

$$C_t = 120 + 120$$

Total capacitance:

$$C_t = 240 \ \mu F$$

Safe working voltage

The safe working voltage in any parallel combination where mixed working voltages are incorporated will be aligned to the *lowest* working voltage capacitor. In the example shown, the lowest value is 10 volts.

Electromagnetic induction

The process of a spontaneous electromotive force originating within an unrelated conductor, merely from a variable current generated from a physically detached and independent source is known as *electromagnetic induction*.

Figure 6.46 illustrates this concept graphically.

Magnetic field

Passing an electric current through a conductor generates a uniform magnetic field around and at right angles to it. The strength of the magnetic field is directly proportional to the current flowing

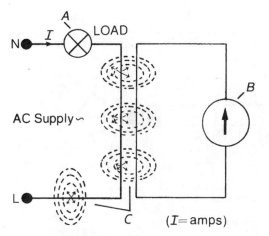

Figure 6.46. Electromagnetic induction. (A, current consuming load; B, galvanometer; C, electromagnetic flux.)

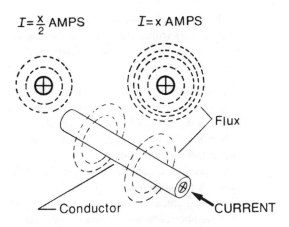

Figure 6.47. The strength of the magnetic field is directly proportional to the current flowing.

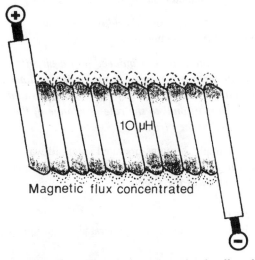

Figure 6.48. Coiling a single conductor has the effect of concentrating the magnetic field.

(Figure 6.47) but it can be increased by coiling the conductor to form a soleniod (Figure 6.48). (The unit '10 μH' used in this figure is explained under 'Unit of inductance' on p. 151.) This has the effect of concentrating the magnetic field by drawing the conductors closer together. A further increase in field strength can be achieved by spiralling the conductor around a ferromagnetic laminated bar.

Generating an electromagnetic force

In 1831 *Faraday* discovered that when a bar magnet was plunged into a simple air-type soleniod

(Figure 6.49), an *electromotive force* (e.m.f.) was naturally generated within the soleniod. He found that the magnitude of the e.m.f. generated was dependent on the following criteria:

1. The strength of the magnetic bar.
2. The magnetic lines of force (flux) must cut through the solenoid at *right angles*. Alternatively, the magnet may be static while the solenoid moves.
3. The speed at which the magnetic flux cuts through the solenoid at right angles.
4. Type of materials used.
5. The number of turns of wire serving the solenoid. (The induced e.m.f. is proportional to the square of the number of turns of wire forming the coil.)
6. The diameter of the solenoid.

From these observations *Faraday* was able to formulate the following law:

An electromotive force is induced into a circuit whenever there is a change in magnetic flux linked to the circuit. The strength of the e.m.f. generated is directly proportional to the rate of change of the flux related to the circuit.

If, for example, a mains-operated solenoid was placed in a circuit the rate of change would equal the frequency of the supply.

Lenz's law

The lines of magnetic flux generated by an induced

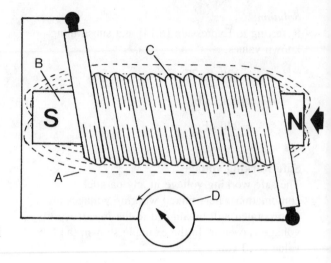

Figure 6.49. An e.m.f. is generated when a bar magnet is plunged into a simple air-type solenoid. Plunging from right to left will cause the galvanometer needle to momentarily flick to the left. (A, magnetic flux; B, bar magnet; C, solenoid; D, galvanometer.)

electromotive force create a magnetic retarding phenomenon by attempting to prevent the actual motion of the current-carrying conductor producing it (Figure 6.50).

If, for example, a current-carrying conductor is rotating in an anticlockwise mode, the induced magnetic field will attempt to motivate the same conductor in a clockwise direction. This concept is known as *Lenz's Law*, named after *Heinrich Lenz*, a German physicist who lived between 1804 and 1865.

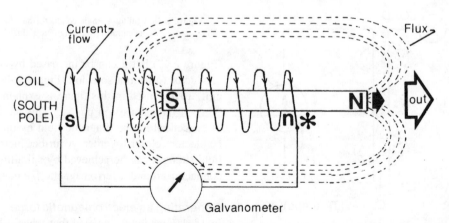

Figure 6.50. When the bar magnet is withdrawn from left to right the galvanometer needle will momentarily flick to the right. ✱ Induced current makes this end of the solenoid a North pole and opposes the movement of the bar magnet.

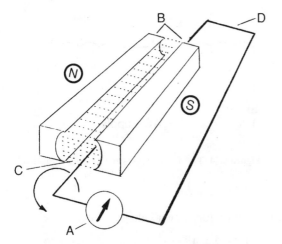

Figure 6.51. When a conductor is drawn across a magnetic field an e.m.f. is induced into it. (A, ballistic galvanometer – an instrument for detecting small electric currents; B, North- and South-seeking bar magnets; C, magnetic flux; D, conductor.)

Fleming's Right-Hand Rule

When a conductor is drawn at right angles across a magnetic field an electromotive force is induced into it. Connecting a sensitive galvanometer in circuit, as illustrated in Figure 6.51, will produce a momentary deflection of the needle. Movement will occur in one direction only, depending on the course the conductor takes across the magnetic field. If, however, the direction of displacement across the field is reversed, a transient current will be witnessed in the opposite direction. This concept is also illustrated as Figures 6.49 and 6.50.

Fleming's Right-Hand Rule is a convenient way to verify the direction of an e.m.f. induced into a conductor as it can symbolically relate the direction of motion, flux and e.m.f. generated.

The thumb, forefinger and second finger of the right hand should be held at right angles to each other. When the forefinger is pointed in the direction of the magnetic flux and the thumb directed to the motion of the conductor, the second or middle finger will indicate the direction of the e.m.f. *Fleming's Right-Hand Rule*, named after *John Ambrose Fleming* (1849–1945), can be studied in graphical form from Figures 6.52 and 6.53.

Unit of inductance

The SI unit of inductance is known as the *henry*, symbol H, self-inductance symbol L, and is named

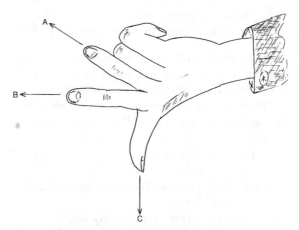

Figure 6.52. Fleming's Right-Hand Rule. (A, direction of e.m.f.; B, magnetic flux, North and South; C, motion of the conductor.)

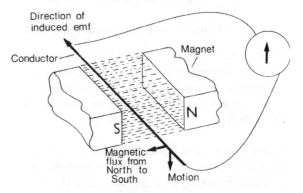

Figure 6.53. Fleming's Right-Hand Rule in schematic form. (Magnetic flux density is measured in *tesla*, defined as the density of one *weber* of magnetic flux per square metre.)

after the American physicist *Joseph Henry* (1797–1878). It is given to be the rate of change of current in a circuit of 1 amp per second which produces an induced electromotive force of 1 volt.

Values of inductors range from about 0.1 microhenry, written as $0.1\,\mu H$, to 10 henries (H). The value of an inductor will depend on the following criteria:

1. Rate of current change (if doubled, then induced e.m.f. doubled).
2. Number of turns of wire.
3. Diameter of coil.
4. Material composition of the wire.
5. Material composition of the core (air, iron or steel but brass to lower induction).

6. Magnetic permeability of the core (this could be referred to as the magnetic factor).

As an example, a 10 mm diameter air-core inductor supporting 100 turns of wire would be far less in value than a similarly sized iron-core component.

The value of inductance in a given circuit may be calculated by use of the following expression:

$$L = \mu_r\,\mu_0\,\frac{aN^2}{l} \text{ henries} \qquad [6.15]$$

where μ_0 is the permeability of free space (1.257 $\times\ 10^{-6}\,\mathrm{H\,m^{-1}}$)

μ_r is the relative permeability of the inductor core

a is the cross-sectional area

l is the length of wire used to form the inductor

L is the unit of self-inductance in henries.

Expression [6.15] provides a brief insight into the depth of problems involving inductive circuits. A complete theoretical and detailed examination is not possible at this stage without straying into another course of study.

Inductors placed in series formation
The total inductance of a series formation may be calculated by adding together the values of the independent inductors. It is the same as evaluating the combined resistance of a group of resistors connected in a smilar fashion. Hence:

$$L_t = L_1 + L_2 + \ldots \qquad [6.16]$$

where L_t is the total value in henries

L_1 is the value of the first inductor

L_2 is the value of the second inductor.

As an example, consider the following problem and Figure 6.54.

Two large ferrite core coils have a mutual inductance of 2 and 4 henries respectively and are wired in series formation as illustrated in Figure 6.54. Evaluate the total inductance of the circuit in henries.

Solution:
Referring to Expression [6.14]:

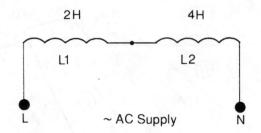

Figure 6.54. Two ferrite coils wired in series formation.

$$L_t = L_1 + L_2$$

Substituting for known values:

$$L_t = 2 + 4$$

$$= 6 \text{ henries}$$

Inductors placed in parallel formation
The total inductance of a parallel group of inductors is the reciprocal of the sum of the reciprocals of the individual inductors. It may be evaluated using the same method as when resolving resistors wired in parallel formation. Hence:

$$\frac{1}{L_t} = \frac{1}{L_1} + \frac{1}{L_2} + \ldots \qquad [6.17]$$

where L_t is the total value in henries

L_1 is the value of the first inductor

L_2 is the value of the second inductor.

As an example, consider the following problem and Figure 6.55:

Two large ferrite core coils have a mutual inductance of 2 and 4 henries respectively and are

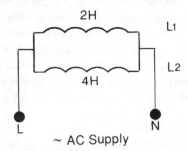

Figure 6.55. Inductors in parallel formation.

TABLE 6.10   Graphical symbols used for architectural and installation plans and diagrams from BS 3939: Section 27, 'Graphical symbols for Electrical, Power, Telecommunications and Electronic diagrams'

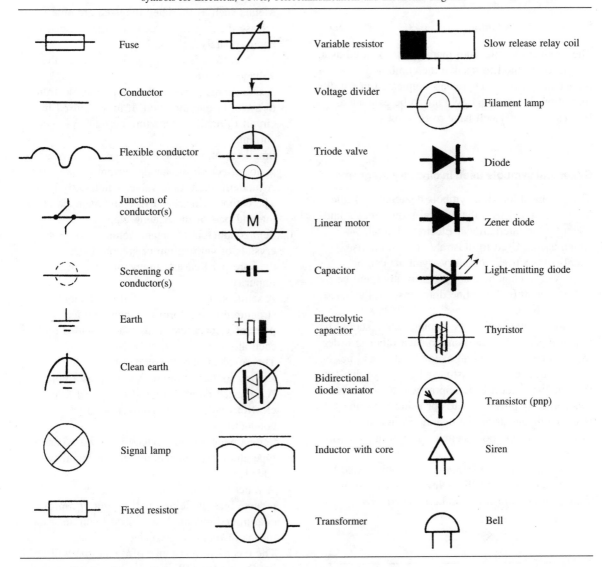

Fuse	Variable resistor	Slow release relay coil
Conductor	Voltage divider	Filament lamp
Flexible conductor	Triode valve	Diode
Junction of conductor(s)	Linear motor	Zener diode
Screening of conductor(s)	Capacitor	Light-emitting diode
Earth	Electrolytic capacitor	Thyristor
Clean earth	Bidirectional diode variator	Transistor (pnp)
Signal lamp	Inductor with core	Siren
Fixed resistor	Transformer	Bell

wired together in parallel formation. Calculate the total inductance of the circuit in henries.

Solution:
Referring to Expression [6.17] and substituting for known values:

$$\frac{1}{L_t} = \frac{1}{2} + \frac{1}{4}$$

First find the lowest common multiple (4):

$$\frac{1}{L_t} = \frac{2+1}{4}$$

$$= \frac{3}{4}$$

Cross-multiply to bring the expression in terms of L_t:

$$3 \times L_t = 4$$

Divide each side of the equation by 3:

$$L_t = \frac{4}{3}$$

$$= 1.333 \text{ henries}$$

The product over the sum method can be used as a simplified method to resolve the combined inductance of two components connected in parallel formation. If more than two are to be evaluated, Expression [6.17] will have to be used.

Graphical symbols used in circuitry diagrams

The graphical location symbols depicted in Table 6.10 correspond with the general requirements laid down by the International Electrotechnical Commission. Used to illustrate and summarise details of electronic circuitry, symbols can be employed in conjunction with each other and be included with BS 3939 line diagrams, as illustrated in Figure 6.56.

Symbols interpreting a common theme should be drawn in similar proportions to each other in order to avoid confusion. As an example, it would be wise to draw a fixed resistor and a voltage divider the same size but a little impractical to schematically present a varistor using the same scale. Components which operate in unison, for example a double pole switch, must be linked with a dotted line.

Always use formal symbols as recommended by British Standards and IEC. Never design and use your own if confusion is to be avoided.

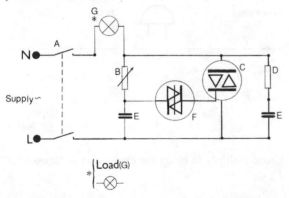

Figure 6.56. A domestic dimmer switch. (A, double pole switch; B, variable resistor; C, varistor; D, resistor; E, capacitor; F, thyristor; G, lighting load.)

Summary

1. Metals such as thoriated tungsten are able to release their surface electrons when heated to approximately 2273 K.
2. Diode current is directly proportional to the voltage applied.
3. A thermionic valve will only permit electrons to travel in one direction. It may be used in circuit to rectify alternating current to direct current.
4. Small variations of grid voltage can have a major effect on the anode current.
5. A modern thermionic valve is indirectly heated. The cathode (the negative component) is not an integral part of the heating element.
6. Semiconductors are manufactured from the crystals of germanium or silicon. Once melted they are purified and impregnated with an impurity.
7. A semiconductor is a crystalline element structured in the form of a cubic lattice sharing pairs of valence electrons; one provided by each atom.
8. The covalent gap, developed by an electron leaving its parent atom, is known as a hole.
9. A balanced combination of holes and free electrons will provide an avenue for conductivity.
10. Semiconductors are divided into two types: n-type and p-type depending on the impurity added.
11. A zener diode is designed to operate in a reverse bias mode and will trigger at a pre-determined potential. It is used in circuit as a means of regulating voltage.
12. The electrical resistance of a material will depend on the following: length, cross-sectional area, temperature, the natural resistivity of the material and, in the case of *selenium*, light.
13. There are many types of resistors. Examples are: carbon based, carbon film, wire wound, metal film, metal oxide and metal glaze.
14. A colour-coded resistor must be read from left to right. *Band 1* represents the first number; *Band 2* the second number; *Band 3* represents the multiplier or the number of zeros following the first two numbers. The last band represents

the tolerance of the resistor as a percentage.

15. A capacitor's dielectric stress is directly proportional to the applied voltage. If the voltage is increased beyond the strength of the dielectric a physical breakdown will occur within the capacitor.

16. There are many different types of capacitors. Examples are: ceramic, silver mica, paper foil, electrolytic and polystyrene.

17. Placing capacitors in parallel formation will enlarge the plate area and therefore increase the value of the combined capacitance.

18. When capacitors are connected in series formation the total capacitance is reduced in value.

19. The process of a spontaneous electromotive force orginating within an unrelated conductor, merely from an alternating current generated from a physically detached and independent source, is known as *electromagnetic induction*.

20. The SI unit of inductance is called the henry, symbol H. The symbol used for self inductance is *L*.

21. In order to obtain a high capacitance, the following criteria are adopted:
    (a) Generous plate area.
    (b) High dielectric constant (also known as permittivity; see Table 6.5).
    (c) Minimum distance between opposing plates.
    (d) Suitable dielectric insulating properties (see Table 6.6).

22. Relative permittivity (the dielectric constant) can be defined as the ratio of capacitance of a given capacitor with a specific dielectric and the capacitance of the same capacitor with only air between the plates. (Symbol ϵ, lowercase Greek letter epsilon.)

23. Resistivity is a constant and represents the resistance offered by a cube of material under review at a temperature of 273.16 K (0 °C). It is expressed in ohms per metre.

24. Silicon- or germanium-based diodes may be tested using a standard multimeter switched to the ohms scale.

25. Light-emitting diodes, used as visual indicators, are made from materials such as gallium arsenide. Wiring is carried out in forward bias and light is emitted at the p–n junction whenever 'holes' and electrons recombine.

Review questions

1. Explain the term 'thermionic emission'.
2. List how electron flow is governed in a simple thermionic diode valve.
3. Specify two materials from which semiconductors are forged.
4. What type of material is produced when a semiconductor is doped with either gallium, boron or indium?
5. The zener diode may be used
   (a) as a visual indicator
   (b) to regulate voltage
   (c) as a smoothing device
   (d) as a means to rectify alternating current.
6. List three types of commercial resistors.
7. Name three external factors which would bring about change in a carbon resistor.
8. Calculate the total resistance in ohms of three 20 ohm resistors wired in parallel formation.
9. What is the name given to the insulating compound sandwiched between the two plates of a simple capacitor?
10. Confirm the following statements:
    (a) A picofarad, symbol pF, is one-millionth part of a farad. TRUE/FALSE
    (b) A farad is the capacitance developed between two plates when connected to a potential difference of 1 volt and charged with 1 coulomb of electrical energy. TRUE/FALSE
    (c) The modern thermionic valve is heated directly. TRUE/FALSE
    (d) Semiconductors forged from the crystals of germanium and silicon have a negative coefficient of resistance. TRUE/FALSE
11. Calculate the total capacitance of three $20\,\mu F$ capacitors placed in parallel formation.
12. Evaluate the combined capacitance of two $6\,\mu F$ capacitors wired in series formation.
13. Determine the total inductance in henries of three 10 H inductors connected in series formation.
14. Describe briefly how a thermionic triode valve may be used.

15. Explain three practical applications for a light-emitting diode.

The answers to these review questions may be found in Appendix D.

Handy hints

1. Electronic equipment should never be flash tested when carrying out portable appliance testing as damage can be caused to printed circuit boards.
2. Care should be taken to avoid foreign contamination such as oil and grease when soldering components or wires together. Contamination will prevent good conductor contact from being made.
3. Take additional care when cutting small single solid conductors with sharp wire cutters. Target the waste copper to the ground as damage to the eye could result from careless cutting.
4. Digital ohm-meters are easily read but care should be taken to observe the position of the decimal point as mistakes can be accidentally made.
5. *Breadboard* systems are both quick and simple to use and have many advantages over *stripboard* circuit assembly work. Wire and components need only be inserted into holes, held firm with tiny spring contacts. Ideally suited for simple circuits and experimental work.
6. Metal oxide semiconductors are very sensitive to static voltages. Should a handler of such a device be charged with static electricity, the oxide insulating material could easily become damaged. High static voltages can be generated by an operative walking across a nylon carpet when he or she is wearing synthetic soled shoes.
7. A portable appliance tester will deliver a flash test voltage of 1.5 kV, in respect to earth, when applied to *Class 1* equipment and 3 kV when *Class 2* equipment is being tested. Never physically touch the appliance while testing is being carried out.

7 Trunking and cable tray systems

Historical introduction

Early trunking systems, made entirely from wood,
were developed during the late 1880s as a means of
protecting electrical wiring. In those days only
affluent businesses and large houses could afford
this new and novel source of energy. A trunking
installation forged entirely from wood was known
as *cabinet casing*, and is illustrated as Figure 7.1.
Basic woodworking skills were necessary as
vulcanised india-rubber cables were laid in grooved
battens which needed to be cut, measured and
mitred to fit the course of the installation. Cabinet
casing was available in various widths, ranging
from one imperial inch (25.4 mm) to three imperial
inches (76.2 mm); but the overall depth was always
maintained at approximately half of one imperial
inch (12.6 mm).

Joining was made by physically marrying the
conductors together, soldering and applying a black
sticky fabric-based electrical tape to finish the
workpiece. The system was sealed by means of a
decorative thin timber lid tacked into position each
side of the grooved base.

Figure 7.1. Early trunking systems made entirely from wood were known as cabinet casing.

Modern systems

Today we are spoilt with an excellent choice of industrial, commercial and domestic trunking systems to suit the design of the majority of installations. Manufactured from steel or high-impact, self-extinguishing, corrosion-free plastic, PVCu trunking must be made from an insulating material with a thermal combustibility factor rated '*P*' as recommended in British Standards directive 476, Part 5. Photo-copies of this directive are readily obtainable if ordered through a leading public library.

Trunking profiles (Figure 7.2) are manufactured to meet the demands of the installation and are available in the following cross-sections:

- Square (general purpose)
- Rectangular (floor, ceiling and general purpose)
- Triangular (workshop benches, cornice installations)
- Delta (workshop and laboratory installations)
- Angled rectangular (skirting installations)

Manufactured in 3, 4, 5 or 6 metre length, depending on both design and usage, trunking is secured by means of a hinged, screwed or snap-on lid. Single-, twin- or triple-compartmental varieties are available to satisfy the requirement of the installation. This is ideal when categories 1, 2 and 3 circuits are to be wired and are designed to be accommodated separately but within a common trunking. *The Institution of Electrical Engineers*

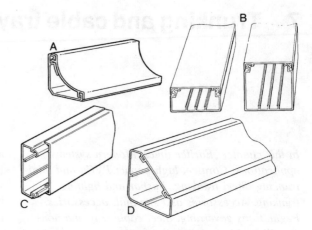

Figure 7.2. Trunking profiles. (A, cornice; B, maxi-trunking; C, dado; D, bench.) (Reproduced by kind permission of *Marshall Tufflex Limited*.)

have categorised circuits in order to avoid confusion and to provide means of recognition.

- *Category 1 circuit*: A circuit operating at low voltage (50 to 600 volts AC) and supplied from the electrical mains.
- *Category 2 circuit*: Any telecommunication circuit, intruder alarm system, data transmission circuit using extra low voltage (not exceeding 50 volts AC or 120 volts DC), and wired separately from other categories.
- *Category 3 circuit*: An emergency lighting or fire detection and alarm circuit.

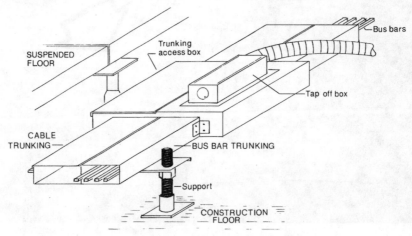

Figure 7.3. Floor access trunking: busbar type.

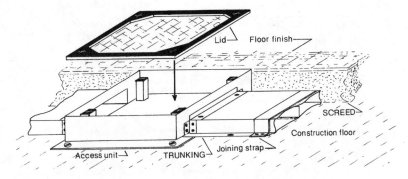

Figure 7.4. Under floor screed trunking.

(Compartmental varieties must be fire resistant when Category 3 circuits are installed.) Wired separately from other categories of circuit. Regulation 528-01-04 confirms.

Designs and applications of trunking

Table 7.1 profiles in general terms trunking types available to suit the design requirements of an installation.

Mini-trunking

Miniature trunking is manufactured from high-impact unplasticised polyvinyl choride, PVCu. Similar in design to its commercial counterpart, only smaller, mini-trunking is available in progressive sizes ranging from 16 × 10 mm to 75 × 16 mm in cross-section as illustrated in Figure 7.5.

Multi-compartment trunking is available in larger sizes (Figure 7.6), many of which are obtainable with self-adhesive backing which is

TABLE 7.1 Trunking design and application

Trunking type	Material used	Design profile	Application/Suitability
Bench	Steel or PVCu	Triangular or delta	Workshops; laboratory benches
Busbar	Steel	Rectangular	Rising mains; cavity floor (see Figure 7.3); industrial use
Busbar	PVCu	Skirting profile	Domestic and commercial use
Cornice	PVCu	Concave or triangular	Used as a domestic cable ducting; can be used with mini-trunking
Dado	PVCu	Rectangular	Domestic and commercial use
Floor trunking	Steel	Rectangular	Industrial and commercial
General purpose	Steel or PVCu	Square or rectangular	Industrial, commercial and general-purpose applications
Lighting trunking	Steel	Square	Lighting installations in industry
Mini-trunking	PVCu	Square or rectangular	Commercial and domestic installations
Skirting	Steel or PVCu	Rectangular	Industrial, commercial and general-purpose applications
Underfloor (screed trunking)	Steel	Rectangular	Used as a general-purpose sealed cable duct as illustrated in Figure 7.4; inspection boxes are fitted where required
Under slab trunking	Steel	Rectangular	Steel trunking is installed beneath the construction slab of a building serving inspection boxes fixed to the floor above; cables are routed from the main trunking system through precast openings to the inspection boxes

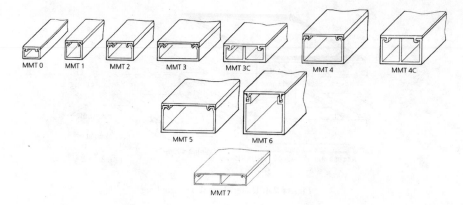

Figure 7.5. Popular sizes of mini-trunking. (Reproduced by kind permission of *Marshall Tufflex Limited*.)

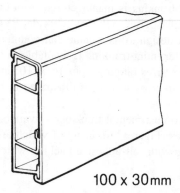

100 x 30 mm

Figure 7.6. Multi-compartmental trunking. (Reproduced by kind permission of *Marshall Tufflex Limited*.)

useful if the installation is carried out where, for example, drilling is forbidden.

The standard colour is white although beige and charcoal are available to order and red is readily obtainable to serve monitored fire alarm installations.

Once wiring has been completed the lid may be snapped on. To remove, the lid may be peeled away from the main body of the trunking by easing the plastic lid with a screw driver at a suitable end. This conforms to the *Wiring Regulations* demanding that a tool is required to remove the lid if screws are not used. This type of trunking system is ideal for domestic and commercial installations or where refurbishment work is required.

Installation techniques
Route planning and decision making must first be considered. Spare sufficient time to reflect on the size of trunking required, then mark out the proposed route, noting where, for example, any 'on-route' right angles, tees, adaptors and stop ends are to be installed. Mini-trunking may be drilled and then fixed by means of suitable pan-headed wood screws placed at 400 mm centres. Alternatively, it may be stuck using a suitable contact adhesive. When self-adhesive mini-trunking is used it is wise to inspect the proposed route in order to confirm that the trunking will effectively bond to the intended surface. If in doubt, reinforce with suitable wood screws. Self-adhesive trunking will not stick to a damp or dusty surface nor will it stick to flaking paint. Be precise when fixing this type of trunking. Make sure that it is correctly positioned and properly aligned Once offered to a good bonding surface it is often difficult to retrieve should a mistake be made. In-line accessories such as angles, tees, adaptors, reducers and stop ends are designed to be snap-fitted to the trunking. For additional strength and security, in-line components may be stuck in place using a suitable PVCu liquid weld. Figure 7.7 depicts a selection of moulded snap-on type miniature trunking fittings.

Thermal expansion
A sensible fixing policy must be adopted whenever miniature trunking is installed. It is wise to drill two supplementary elongated holes, one above the other, positioned 100 mm from the end of each length of trunking fitted. This is in addition to fixing holes drilled throughout its length, illustrated as Figure 7.8. Washers should be placed behind

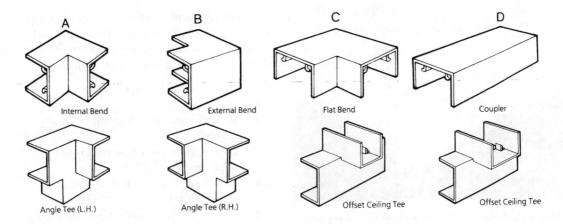

Figure 7.7. Mini-trunking fittings. (A, internal bend; B, external bend; C, flat bend; D, coupler.) (Reproduced by kind permission of *Marshall Tufflex Limited*.)

each pan-headed fixing screw but should not be over-tightened. This will allow freedom for expansion and reduce the risk of distorting the trunking.

All butted joints should be made secure with an internal coupling. These are either stuck on or snapped on, depending on the design. On very long runs of straight trunking an internal expansion coupler should be fitted between every other length of trunking as depicted in Figure 7.9. This is simply done by cementing one end of the expansion coupler permanently to the interface of the trunking while the other end is coated with a non-hardening mastic to help provide the required degree of protection against unsightly buckling due to thermal expansion.

Cornice trunking

Cornice trunking, made from high impact PVCu has been specially designed to harmonize at the intersection point of both ceiling and wall. Installations carried out using this method are usually straightforward providing mistakes made when measuring are not followed through to the cutting process. All fittings slightly overlap at the joints enabling the work to be carried out to a high standard. Cornice trunking manufacturers such as *Marshall Tufflex* often design their product to be compatible with the smaller sizes of mini-trunking. This is an advantage especially when serving switches or ceiling-mounted luminaires. This type

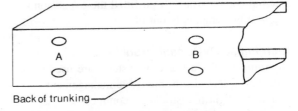

Figure 7.8. Drilling elongated holes will allow freedom of expansion and will reduce the risk of distorting the trunking. (A, supplementary elongated holes; B, fixing holes.)

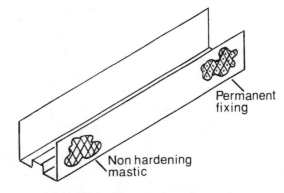

Figure 7.9. A mini-trunking expansion coupler.

of cable management system is supplied with cable retainers which can be increased in numbers should it become necessary. Figure 7.10 illustrates a typical cornice trunking arrangement. On long runs it is wise to allow a small gap in order to accommodate expansion. Cornice trunking is made in 3 metre lengths so a 3 to 4 mm gap should be allowed

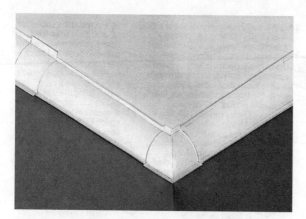

Figure 7.10. Cornice trunking. (Reproduced by kind permission of *Marshall Tufflex Limited*.)

between each length installed. Once both lid and external coupling have been fitted the expansion gap will be completely out of view.

High-impact PVCu cable trunking

High-impact PVCu trunking systems are used to provide a safe and sensible cable management system for light industrial and commercial installations where it might be impractical to use conduit as a means of distribution. Although not as strong as steel and unable to be used as a current

protective conductor, PVCu trunking has many practical advantages over its steel counterpart.

Listed in random order are points to consider when undertaking an installation such as this:

- Both light and easy to handle
- Quicker to install
- Risk of minor personal injury greatly reduced
- High impact and conforms to the requirements of BS 4678, Part 4
- Size range from 50 mm × 50 mm to approximately 150 mm × 150 mm.
- Can be selected with internal division for cable segregation.
- Variety of trunking accessories in the form of tee pieces, angles, couplings and reducers (see Figure 7.11).
- Accessories to match the trunking are designed to be either stuck or pinned together with purpose-made rivets.
- Coloured mainly white but grey is also available although not in all ranges produced.
- Clip-on lids may be staggered over the trunking joints to provide additional strength to the system.

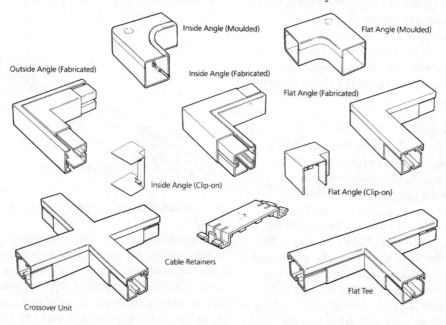

Figure 7.11. PVCu cable trunking fittings. (Reproduced by kind permission of *Marshall Tufflex Limited*.)

- Lids, designed to be snapped on, may be easily removed by peeling back from a suitable end. Alternatively, locking screws or clip-fixers may be employed to secure the lid.
- Snap-on lids can be efficiently fixed in place by securing the first 40 mm or so at a convenient end and ranging the remainder of the unfixed lid at an angle of about 30° to the trunking. With one hand holding the lid and the other pressing the lid onto the trunking it can be quickly put into place. In difficult situations use a mallet or a rubber hammer to gently tap the lid into place.
- Expansion can be accommodated by allowing a small gap at each join.
- Aluminium screening sections may be fitted to protect sensitive circuits from external interference.

Steel cable trunking

This type of cable management is used in industrial situations. First determine the number of cables to be installed allowing due regard for future wiring. Once the number has been calculated aim for a space factor of 45 per cent.

Cables bundled into trunking to the extent of having to compress them in order to secure the lid could lead to serious mechanical and thermal problems. It can be tempting to over-accommodate a trunking installation especially if additional unexpected circuits are not planned for. However, it is far easier to cater for future wiring by planning a generous space factor.

When both size and type of trunking have been decided, route planning can commence by marking the positions of all horizontal and vertical tees, reducers, inside and outside bends and accessories to be used in conjunction with the installation. A chalk-line is a very useful tool when fitting lengths of horizontal trunking. When very long lengths of surface wall-mounted trunking are needed it is quite difficult to maintain a perfectly straight run. In cases like this, an *industrial laser* or stretched chalk-line can help with alignment.

Tools

Unlike a mineral-insulated cable installation, no specialized tools are required for assembling steel trunking. Other than a jigsaw (a machine fretsaw), the majority of tools form part of an average electrician's tool kit. Listed is a selection of basic tools required to carry out the job efficiently:

1. Hacksaw
2. Junior hacksaw
3. File medium cut
4. Retractable rule
5. Mitre block
6. Jigsaw
7. Set square: adjustable
8. Electric drill (110 volts): drill bits
9. Pencil: soft grade
10. Chalk-line
11. Water level: 300–600 mm in length
12. Masonry drill bits
13. Suitable screwdrivers
14. Variety of hole cutters

Installation points

Listed, are installation points which will be of help when new found skills are applied to a steel trunking system:

1. Standard steel cable trunking is usually supplied in 3 metre lengths but lighting trunking is manufactured in 3.6 or 4.5 metre lengths. For general use, two finishes are available: stove enamel on zinc-coated steel for dry conditions and galvanised steel for damp or humid conditions. For very special installations a stainless steel finish is available.
2. Fittings must complement the main body of trunking.
3. Manufacturers will make steel trunking systems to order.
4. Multi-compartmental trunking is obtainable to suit installation needs (Figure 7.12). Dividing fillets are usually spot welded to the base.
5. Optional internal dividing fillets can be provided to fit standard trunking in order to cater for the segregation of cables. These are fitted with small self-tapping screws or, alternatively, nuts and bolts.
6. Trunking lid can be removed by coin-in-slot

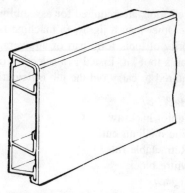

Figure 7.12. Multi-compartmental trunking. (Reproduced by kind permission of *Marshall Tufflex Limited*.)

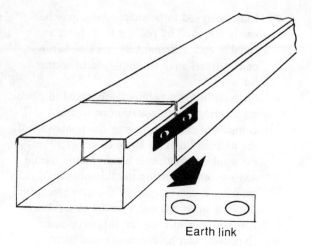

Figure 7.14. To provide adequate means of bonding, copper earth links must physically join each length of steel trunking.

stud; machine fixing screws or turnbuckles. Snap-on lid is usually provided for lighting trunking. This complies with the requirements of the *Wiring Regulations*.

7. Steel cable-trunking arrangements must be completely erected before cables are laid. Regulation 522–08–2 confirms.

8. Measure the length to be installed. Use a soft grade pencil and aided with a tee square inscribe around the periphery of the trunking. Cut with a medium grade hacksaw blade keeping to the waste side of the trunking (Figure 7.13). This will ensure that the measured workpiece is cut to the intended length.

9. Trunking may be secured to a wall or ceiling with pan or round-headed screws. Alternatively, fix to site fabricated or

10. When installed vertically, trunking must have steel *cable retaining straps* fitted.

11. Copper *earth links* must physically join each length of steel trunking to provide adequate means of bonding. Figure 7.14 illustrates.

12. Avoid torsional (*twisting at one end while the other is held*) and tensile stress (*stretched under own weight*) when installing cable as this will lead to problems in the future. Regulation 522–08–06 confirms.

13. Avoid tight bends in the cables as confirmed by Regulation 522–08–03. Always allow a good radius when negotiating trunking tees and angles.

14. When cables installed have a lower thermal rating than the majority, all cables must then be regarded as having a thermal rating which is equal to the lowest rating. Regulation 521–07–03 confirms.

15. Fire barriers, where necessary are fitted after all cables have been laid. Glass wool should be packed into the trunking at a point where the barrier is to be and left proud of the top of the trunking. Once the lid is secured the glass wool will become compressed and provide an adequate barrier against the spread of fire.

16. Care must be taken when fixing the lid to

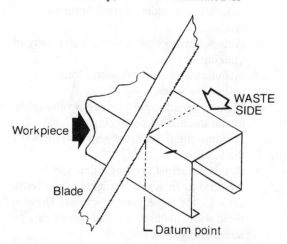

Figure 7.13. Keep the cut to the waste side of the workpiece.

Figure 7.15. Trunking and conduit continuity tester. (Reproduced by kind permission of *Robin Electronics Limited*.)

ensure that cables are not trapped, pinched or scuffed between the lid and main body of the trunking.

17. Before commissioning, test the continuity of the steel trunking by use of an *analog or digital loop tester*. It is important that a very low resistance value is recorded as the main body of the trunking system also acts as the principal current protective conductor. If a problem is highlighted a secondary test may be carried out using a *conduit/trunking continuity and impedance tester*. This test must be carried out with extreme care as a 25 amp low-voltage test current supplied by the instrument is injected into the installation (Figure 7.15). The impedance value in ohms is then read from the analog or digital display panel.

18. The maximum distance between supports for cable trunking can be taken from *Table 4D* in the *IEE's On Site Guide*.

19. Finally, paint with an appropriate finish any damage caused to the trunking during the course of erection. This will guard against corrosion and also make the installation more presentable.

Busbar trunking

Floor trunking

Floor busbar trunking is often a useful away to distribute power when, for practical or aesthetic reasons, an overhead system would not be appropriate. Available to serve both single or three-phase systems, a typical floor trunking busbar arrangement would be rated to provide 63 amps. The system is designed enabling fused tap-off points to be installed wherever required throughout the installation.

Methods of installation

Components and accessories are '*handed*' to ensure that correct polarity is maintained throughout the construction phase of the installation. Fitting one way only guarantees that all component parts are assembled correctly.

Laying floor trunking will require the cooperation of the main building contractor to provide datum lines and measured concrete dunes so that the completed installation will be level with the finished floor screed. Figure 7.16 will illustrate this concept more clearly.

When the builder has provided the information required, a long 'straight edge' together with a 600 mm water level present sufficient means to

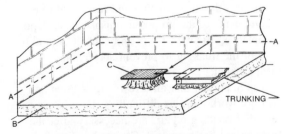

Figure 7.16. Concrete dunes, and datum lines supplied by the main contractor, must first be prepared before floor trunking is laid. This ensures that the installation will be level with the finished floor screed. (A, datum line showing level of finished floor; B, construction floor level; C, concrete dune capped with slate.)

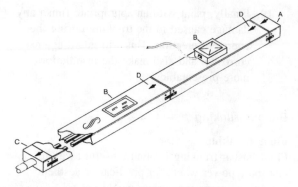

Figure 7.17. A track stop-end is fitted when the floor trunking is terminated. (A, track stop-end; B, power tap-off point; C, cable supply termination unit; D, polarity arrow markings.)

maintain the correct floor trunking levels throughout the installation. It is an advantage to have many datum lines and points of reference within the area to be worked as this will enable the completed trunking to harmonise more easily with the finished floor. Assembling is similar to working with steel box trunking. A retractable rule, soft pencil, large set square and a reliable hacksaw are usually all the hand tools required. Once measured, cut and drilled for coupling, a track interconnection unit is installed to provide busbar continuity. Upon completion the system is supplied with suitably sized cables, often 16 mm², terminating in a track supply service cable box. When floor trunking is terminated, a track stop-end box is installed as Figure 7.17 graphically depicts.

On completion

On completion the general foreman or site manager should be asked to inspect the level of the trunking to confirm that it is suitable and meets with the requirements of the proposed finished floor. Upon acceptance, it is wise to fix and cement around the base of the trunking to prevent accidental displacement.

It is important that all screws are secured and will not work loose as difficulties can arise when remedial bonding repairs are necessary after the floor screed has been laid. Copper bonding links must bridge each trunking joint to provide adequate means of electrical continuity.

Lids or cover plates must always be secured before the floor screed is laid and it is wise to

protect the fixing screws by placing a short length of PVC electrical tape over each screw head. This will effectively prevent screed contamination occurring. Some lids are provided with a protective PVC film overlay which is removable upon commissioning. This should be retained to prevent damage from occurring and only removed when the installation is formally handed over to the client.

Finally, an impedance test must be carried out on the completed installation. This should be implemented using a *conduit trunking continuity and impedance test meter*. If the value obtained is high, an on-site inspection check must be made on all couplings and copper bonding links serving the system. A further test is necessary after remedial work has been effected. (See *Trunking: cross-section area and resistance* later in the chapter.)

Industrial busbar trunking

Industrial, metal clad busbar trunking systems are often designed for use in factories as a means for power distribution to serve machinery. In high-rise commercial buildings the system provides a simple means of power dispensation to each floor (Figure 7.18). It may be selected for use with either single- or three-phase systems and, as with floor trunking, industrial busbar trunking is 'handed' to maintain correct polarity throughout the installation. The busbars are forged from copper but occasionally aluminum is used.

Regulations 522–03–01 to 12 demand that protective measures appropriate to external influences must always be considered. Trunking finishes may be chosen from stove enamel, galvanised or totally insulated to meet the conditions of the installation.

Supporting accessories in the form of angles, bends, tees and 'tap-off' boxes are readily obtainable. Busbar trunking is rated up to 300 amps in sectional lengths between 1.5 and 1.7 metres. Fused tap-off units can be incorporated wherever necessary to provide power and general lighting arrangements. Regulation 523–03–01 requires fused tap-off points for use within an installation.

Using steel trunking as a protective conductor

As the exposed conductive part of a steel trunking installation is used as a protective conductor,

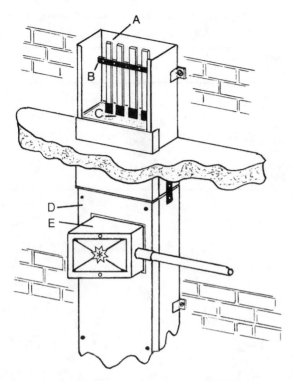

Figure 7.18. Busbar trunking used for power distribution. (A, copper busbars; B, insulated busbar supports; C, busbar sleeving; D, removable cover; E, power tap-off box.)

continuity must be guaranteed throughout its length. Sectional butted joints must be electrically bonded across each joint by means of copper links maintained throughout the length of the trunking installation.

Regulation 543–01 requires the cross-sectional area of a steel trunking arrangement, which forms the protective conductor, to be equivalent to that

resulting from the application of the following expression:

$$S = \sqrt{\frac{I^2 \times t}{k}} \, \text{mm}^2 \qquad [7.1]$$

where S is the cross-sectional area in mm^2
  I is the value of the fault current in amps which would flow through the associated protective device (this could be a semi-enclosed fuse, MCB, etc.)
  t is the operating time in seconds for the protective device to disconnect a circuit sustaining a fault condition
  k is a factor dependent on the material composition of the protective conductor; its initial and final temperatures and the type of insulation serving the protective conductor.

Resolving the value of I: This requires knowledge of the impedance of the supply transformer together with the impedance of both phase and neutral conductors serving the consumer's switchgear. I can then be evaluated by dividing the phase-to-earth voltage by the given impedance.

Resolving the value of t: This may be obtained by referring to the technical characteristics printed on the protective device

Resolving the value of k: This is formulated on an assumed initial temperature, a final temperature during a fault condition and the material composition of the protective conductor together with the type of insulation afforded. Values of k may be found in *Table 43A* of the *Wiring*

TABLE 7.2   Calculating the effects of a fault current

Values of k for general-purpose cables and suitable for disconnection times up to 5 seconds

Material composition of the conductor or type of cable	Insulating material	Initial and final assumed temperatures (°C)	Value of k
Copper up to 300 mm^2 cross-sectional area	PVC (70 °C)	70/160	115
	PVC (85 °C)	85/160	104
	Rubber (60 °C)	60/200	141
MICC plastic covered	Magnesium oxide	70/160	115
Aluminium up to 300 mm^2 cross-sectional area	PVC (70°C)	70/160	76
	PVC (85 °C)	85/160	69

Regulations or referred to in abridged form as Table 7.2. Expression [7.1] is only relevant for disconnection times of five seconds or less.

Cross-sectional area and resistance

In Britain, manufacturers of steel trunking and accessories must conform to British Standards 4568 and 4678. Briefly this requires that the electrical resistance offered does not exceed 0.005 ohm per metre.

Listed in Table 7.3 is a selection of nominal cross-sectional areas for steel trunking. These are modelled on approved sizes published in BS 4678.

Providing the system is installed correctly and all mechanical joints are sound, steel trunking will serve as an excellent protective conductor as Table 7.3 will show.

The *Wiring Regulations* requires a formal test to be carried out to verify that all exposed conductive parts forming the main body of the trunking system will offer a low impedance path to earth.

As a rough guide to the value which may be expected from such a test, the following expression may be used:

$$R = m \times K_r \qquad [7.2]$$

where m is the total length in metres of the
trunking under test
K_r is the factor 0.005 ohm per metre taken
from BS 4568
R is the resistance in ohms.

As a practical example consider Figure 7.19 and the following:

A 52.5 metre length of 100 mm × 50 mm steel trunking is offered for test. Given that the maximum resistance of the trunking conforms to BS 4568, that is, 0.005 ohm per metre, calculate the maximum pratical value that can be expected.

$$\text{Resistance } (R) = m \times K_r \quad (\text{Expression } [7.2])$$

$$R = 52.5 \times 0.005$$

$$= 0.2625 \text{ ohm}$$

If the evaluated resistance is very much less than that obtained by instrumentation, serious

TABLE 7.3   Cross-sectional area of steel trunking

Reference is made to BS 4678

Steel trunking (mm)	Cross-sectional area (mm^2)
50 × 37.5	125
50 × 50	150
75 × 50	225
75 × 75	285
100 × 50	260
100 × 75	320
100 × 100	440
150 × 50	380

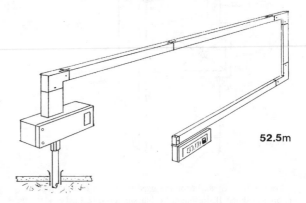

Figure 7.19. A 52.5 metre, 100 mm × 50 mm, steel trunking installation offered for test.

consideration must be given to each trunking coupling installed. A check should be made to ensure that each joint is sound, tight and is fitted with a copper link bridging the butted union.

Fire barriers for steel trunking systems

Suitable fire barriers must be provided around 'rising main' busbar trunking where the installation passes between floors and walls in order to prevent the spread of fire. Holes and gaps left around the trunking must be filled using the same level of fire protection as the surrounding infrastructure.

Cables serving or leaving the busbar trunking must be thermally rated for the maximum working temperature of the busbars. Regulation 523-03-01 confirms.

Lighting busbar trunking

Plug-in busbar trunking for general lighting arrangements is used in commercial, industrial and

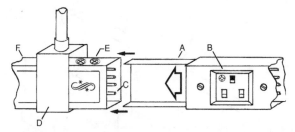

Figure 7.20. Lengths of busbar lighting trunking may be both mechanically and electrically jointed with a single action and secured with captive screws. (A, trunking coupler; B, socket take-off point; C, busbars; D, trunking support attached to ceiling; E, captive screws; F, lighting trunking.)

agricultural buildings where the call for an extensive switching schedule is unwarranted. This is an ideal lighting system for hypermarkets, factories and workshops where a good level of lighting is required.

Current rated between 25 and 40 amps, plug-in busbar lighting trunking is available with from two to six conductors. As with other steel trunking systems, the metal infrastucture is used as a current protective conductor.

Lengths are between 3 and 5 metres long and are reasonably lightweight, so easy to handle. With fixing brackets firmly installed, straight lengths of trunking can be offered up and locked into position, as illustrated in Figure 7.20. The integral busbar arrangement is connected by snapping each length into position and securing by means of captive screws. The electrical connection to the luminaire is made using a standard 13 amp plug fused appropriately.

PVCu skirting and dado busbar trunking

PVCu skirting and dado busbar trunking systems are found in hospitals, schools, colleges, offices and commercial premises where cable management is paramount.

This system is generally multi-compartmental, each compartment supporting its own independent lid. Figure 7.21 illustrates a section of *Sterling Busbar Trunking* manufactured by *Marshall Tufflex* of Hastings, England. One of the thoughtful safety features is an internal busbar cover which prevents accidental direct contact to live conductive parts.

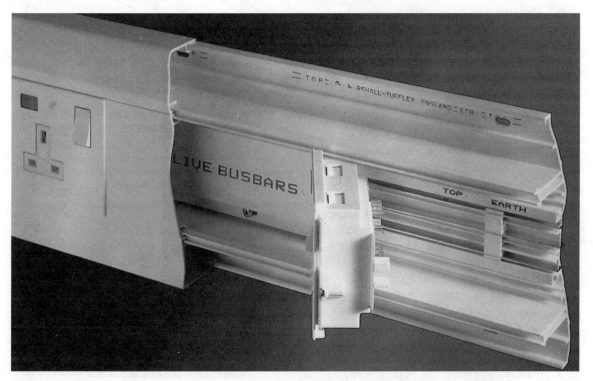

Figure 7.21. Sterling busbar trunking. (Reproduced by kind permission of *Marshall Tufflex Limited*.)

An optional range of electrical accessories and trunking components are available enabling a good standard of work to be carried out.

Site-built trunking accessories

Trunking right angle : internal bend

Designer made bends and sets can be made on site whenever manufactured accessories are unavailable. With practice, they can be quickly fashioned and assembled in a straightforward manner using the tools and equipment listed below:

- Hacksaw (*25–30 teeth per 25 mm of blade*)
- Junior hacksaw or jigsaw
- Retractable rule
- Adjustable steel set square
- Soft pencil or fine line permanent marker pen
- Battery-operated drilling machine or a 110 volt drill
- Selection of hole cutters and drill bits
- File: medium cut
- Pop rivet gun (*if available*).

Construction guidelines

The following paragraphs describe how a simple 90° right-angled bend can be designed and assembled under site conditions.

1. Select a short section of trunking between 900 and 1000 mm in length. Using a soft pencil, aided with a reliable set square, draw a line around the periphery of the trunking. This should be done at a mid-point position as described in Figure 7.22.
2. Check the *exact* diameter of the workpiece and transfer this measurement to either the top left- or right-hand side of the central datum line as depicted in Figure 7.23. Punctuate with a mark. Using an adjustable set square as a guide, draw a pencil line from the marked trunking to the bottom of the centre datum line as shown in Figure 7.24. Repeat this guideline on the opposite vertical side.
3. At this stage there will be a right-angled triangle drawn on each outer side of the

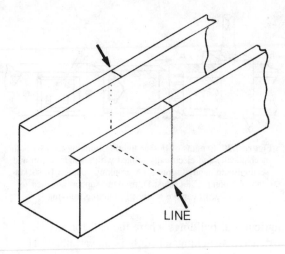

Figure 7.22. Draw a line around each side of the trunking.

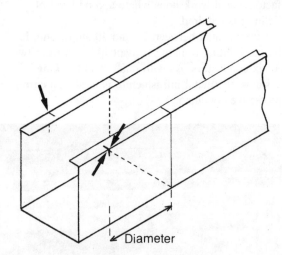

Figure 7.23. Measure the diameter of the workpiece and transfer this measurement to either the top left or right side of the datum line as shown in Figure 7.22.

trunking. Remove the two triangles using a hacksaw, making sure that the cut is ranged to the 'waste' side of the measured lines. Once removed, file off all jagged or rough edges in order to protect future cables from damage. The trunking will now appear as illustrated in Figure 7.25.

4. Cut a portion of surplus timber with a good square edge on one side able to be fitted comfortably across the internal diameter of the trunking. Range the wooden template to

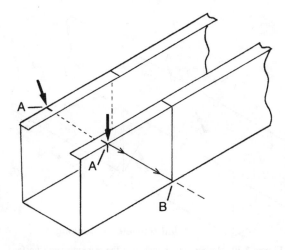

Figure 7.24. Draw a pencil line from the marked trunking to the bottom of the centre datum line on both sides. (A, marked trunking; B, bottom of datum line.)

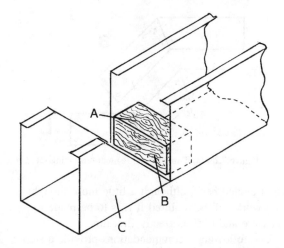

Figure 7.26. Place the prepared timber to the side adjacent to the vertically cut side. (A, wooden template; B, square edge of timber; C, steel trunking.)

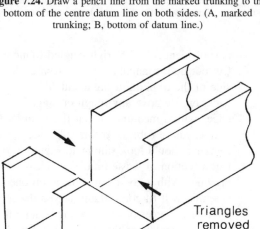

Figure 7.25. Remove the two pencilled triangles with a hacksaw.

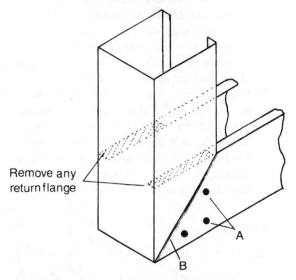

Figure 7.27. Bend into shape. Check for squareness and fortify with pop rivets. (A, pop rivets or nuts and bolts; B, angled cut.)

the side adjacent to the vertically cut side as shown in Figure 7.26. Hold the timber firmly in place, the square edge ranged next to the proposed bending point, and with the other hand push up the side of the trunking adjoining the angled cut. Allow the vertically cut sides to be sandwiched between the angled trunking sides as Figure 7.27 illustrates. The timber insert will help to provide a sharper edge at the bending point and consequently be far more presentable to the eye. Once completed,

dress the bend with a hammer and remove the wood.

5. Check for squareness and fortify with pop rivets or suitably sized nuts and bolts as out lined in Figure 7.27.

Trunking sets and return sets

A trunking set or return set (Figure 7.28) is constructed in a similar fashion as when making a

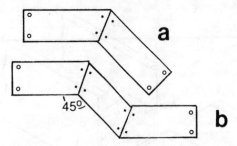

Figure 7.28. (a) Trunking set; (b) return trunking set.

right-angled bend, although a little more thought and work will be required if it is to be made to measure and to fit correctly first time.

The following recommendations provide a step-by-step guide for the operative who is building this component for the first time.

1. Select a section of trunking that can be worked comfortably yet is long enough to accommodate either a set or return set. Draw a datum line, using a soft pencil and set square as a guide, around the three sides of the trunking as shown in Figure 7.29.
2. Verify the *exact* diameter of the workpiece and transfer *half* of this measurement to the top left or right hand of the centre datum line of the trunking. After punctuating with a suitable mark, draw a line to the bottom of the vertical datum as shown in Figure 7.30. Repeat on the opposite side to produce a

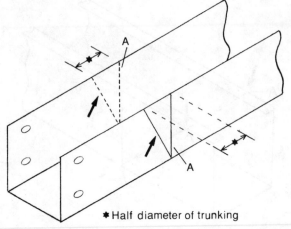

Figure 7.30. Measure the diameter of the trunking and transfer half of this measurement to either the top left- or right-hand side of the datum line. Draw a line from this mark to the base of the datum line, A.

pair of identical 22.5° right-angled triangles,

3. Cut out both triangular shapes from each side of the trunking using a suitable hacksaw. Remove any rough or jagged edges with a medium cut flat file in order to protect future cables. At this stage the trunking should look similar to Figure 7.31.
4. Cut a section of waste timber which is provided with a good square edge on one side that will fit comfortably across the internal diameter of the trunking. Range the wooden template to the side adjacent to the

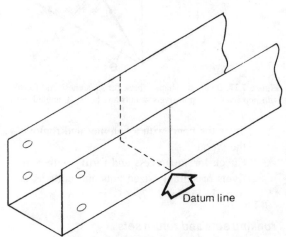

Figure 7.29. Draw a datum line using a soft pencil, guided by a set square, around the three sides of the trunking.

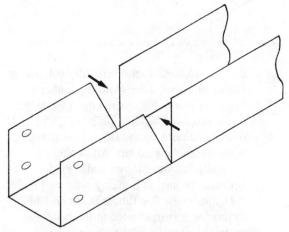

Figure 7.31. Cut both pencilled triangular shapes from each side of the trunking.

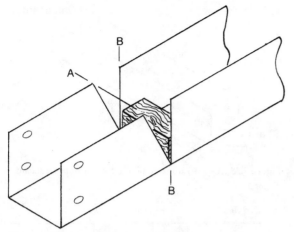

Figure 7.32. Place the wooden shape to the side adjacent to the vertically cut datum line. (A, wooden template; B, vertically cut trunking.)

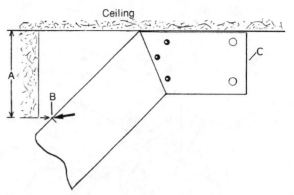

Figure 7.34. Measuring the depth of the required set. (A, required depth of set; B, point of set; C, shortest leg.)

vertically cut datum line as shown in Figure 7.32. Secure the timber firmly with one hand making sure that the squared edge is ranged next to the bending point while gently bending up the remaining half of the workpiece until a 45° set is formed. The timber insert will help to provide a good sharp ridge at the bending point and will improve the quality of the set by making it more acceptable to the eye. Allow the vertically cut sides to be sandwiched between the angled trunking sides.

5. Confirm that the angle is correct and secure using pop rivets or, alternatively, nuts and bolts. Figure 7.33 represents a very simple

set. If a return set is required, additional work must be carried out.

6. *Measuring and preparing the depth of a return set*: Draw a line of reference on a suitable flat surface and offer the shortest section of the set to the line as described in Figure 7.34. Measure the depth required and mark the trunking. This will now become the point at which the workpiece is formed into a returned set.

7. Prepare the trunking as shown in Figure 7.35.

8. *Cutting to shape and forming, the return set*: Cut out the *complete* left-hand side of the centre line comprising two side triangles bridged by a rectangular base section. To assist with electrical continuity and mechanical stability, the trunking lip must

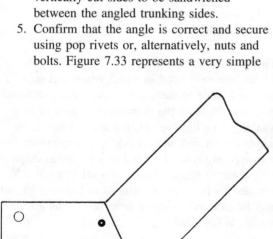

Figure 7.33. Bend into shape. Fortify with pop rivets.

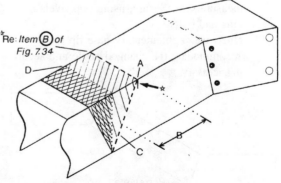

Figure 7.35. Preparing the returned set. (A, 'depth of required set' mark; B, same diameter of trunking; C, reference line; D, centre line.)

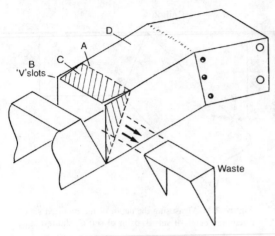

Figure 7.36. A small 'V' should be cut along both sides of the bottom edge of the trunking from the vertical centre line to the angled dotted line. (A, 'depth of set' mark; B, 'V' slots; C, supplementary lip; D, bottom of trunking.)

remain unbroken. A 2 mm diameter 'V' shape should be cut along, both sides of the bottom edge of the trunking from the vertical centre line to the angled dotted line as depicted in Figure 7.36. This will act as a supplementary lip and help stablise the return set when it is assembled.

9. Gently bend the workpiece and unite both sections of the return set allowing the vertically cut centre edges (which once formed the centre line) to be sandwiched between the angled sides of the trunking as illustrated in Figure 7.37.

10. Check that all angles are correct and that the set has been worked to meet the required measurement. Secure, using pop rivets or nuts and bolts.

11. Dress the supplementary base lip to accommodate the changed angle and secure if necessary.

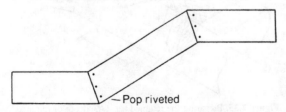

Figure 7.37. A site-constructed return set.

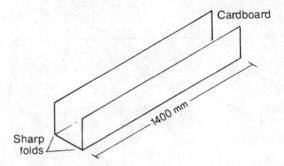

Figure 7.38. A model trunking profile formed from cardboard.

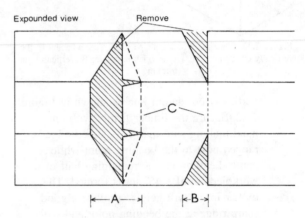

Figure 7.39. Cut to shape with a pair of scissors or safety razor. (A, diameter of trunking; B, half diameter of trunking; C, bend points.)

12. Fit copper continuity earth links where required.

The way to perfection

An alternative method in which return sets may be practised while in the comfort and privacy of home is to make a small but manageable length of mini-trunking from lightweight cardboard. This can be accurately formed into a model trunking profile as portrayed in Figure 7.38 and may be cut to shape with a pair of sharp scissors or a safety razor. The returned set is shown as detailed in Figure 7.39 and may be used as a point of reference if this learning method is favoured.

Cutting rectangular or square holes in trunking
Figure 7.40 illustrates a typical installation served by a metal clad distribution centre butted to the top side of a steel trunking arrangement by way of

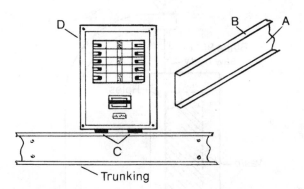

Figure 7.40. Leaving a space between the top of the trunking and distribution centre will allow the covers to be fitted without the need to remove a section of lip. (A, trunking cover; B, lip of cover; C, spacers; D, distribution centre.)

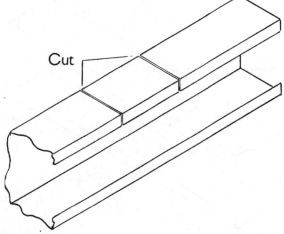

Figure 7.42. Draw two measured parallel lines and cut with a hacksaw to the rear of the trunking.

nuts, bolts and spacers. Provided a margin is allowed between the trunking and distribution centre, the cover can be fitted without having to remove a section of top lip. Retaining the lip will also provide additional durability.

Spacing devices can either be designer made from flat-iron and drilled or, alternatively a 20 mm diameter steel conduit locking ring may be used (Figure 7.41). Whichever method is adopted, both trunking and distribution centre must be drilled and bolted to each other to contribute towards electrical continuity. Provided the depth of the distribution centre is greater than the horizonal width of the trunking, a side portion of trunking may be removed in order to provide access for cables leaving the distribution centre. This is considered more desirable than drilling holes in both trunking and distribution centre. Cutting holes is both time consuming and costly, as 'male' type brass bushes

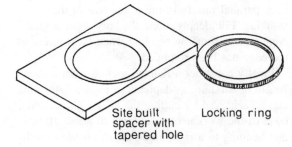

Site built
spacer with
tapered hole   Locking ring

Figure 7.41. Spacers can either be designer made or a 20 mm diameter conduit locking ring may be used.

and locking rings must be fitted to protect the cables from damage.

Removing a side section of trunking

The majority of fuse boards and distribution centres now have removable retangular plates fitted to both top and bottom with weakened 'knockouts' of various sizes.

Remove either top or bottom plate and measure the exact length of the rectangular hole, transferring this dimension to the side of the trunking. Draw two parallel lines accommodating the measurement taken and cut to the rear of the trunking as outlined in Figure 7.42. Select a good sharp medium cut file and rasp along the back right-angled edge. When about 50 per cent of the metal is removed the worked section can be easily snapped off by bending backwards and forwards. Dress all rough and jagged edges with a file and add plastic or lead edging strip to finish off. Remember to clamp a suitably sized earthing conductor from the main body of the trunking to the principal earthing terminal in the distribution centre, to provide additional continuity.

Trunking slots

Slots, providing access for cables, can be shaped by applying the following guidelines:

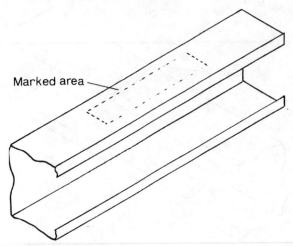

Figure 7.43. Decide on the shape of slot required and mark accordingly.

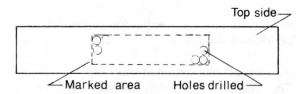

Figure 7.44. Drill two sets of holes in opposite corners of the workpiece.

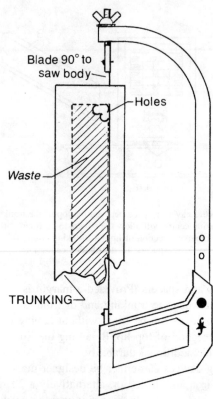

Figure 7.45. Align the blade of the hacksaw to accommodate the prepared holes and carefully cut the required shape.

1. Decide on the shape of the area to be removed and mark the trunking accordingly, (Figure 7.43).
2. Drill two sets of 6 mm diameter holes in opposite corners as illustrated in Figure 7.44. This can be achieved by careful measurement and a sharp metal punch enabling the holes to be drilled within the waste area of the trunking.
3. Realign the blade of the hacksaw to accommodate the prepared holes at an angle of 90° to the main body of the saw, as described in Figure 7.45.
4. Carefully cut the shape required, adjusting the alignment of the hacksaw whenever necessary. A jigsaw will make the task far easier.
5. Dress the exposed edge with a file to remove roughness and apply plastic or lead edging strip to finish off.

Site-built trunking tee sections

Site-fabricated tees are straightforward to assemble and can be produced to an acceptable standard in a reasonable time.

First select a manageable section of trunking to represent the horizontal member of the tee. Draw two parallel lines from a mid-point position to correspond with the diameter of the vertical leg of the fitting. Cut, using either a hacksaw or jigsaw, to each parallel line finishing at the rear of the trunking. File deeply along the back right-angled edge and snap off by bending when 50 per cent of the edge has been removed.

Finally construct a flanged section from a short length of trunking as depicted in Figure 7.46. This can be achieved by cutting the back right-angled edge of the workpiece to a depth of about 10 mm and bending to a right angle after removing the lip from the trunking. This must be done on both sides as shown in Figure 7.46(b).

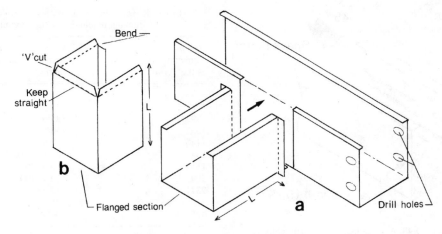

Figure 7.46. Site-made trunking tee section.

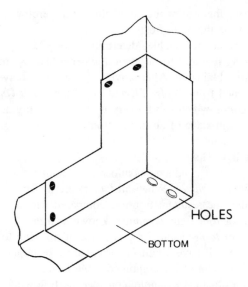

Figure 7.47. As a precaution against accumulated condensation, two or three small holes can be drilled in the trunking at its lowest level.

Once completed, drill and pop-rivet the workpiece. Fit copper bonding links to bridge each leg of the newly formed trunking tee.

Cable trunking systems and the IEE Regulations

Vertical runs (reference is made to Section 422)

Internal heat barriers must be provided in vertical trunking runs to prevent the ambient temperature at the top of the trunking from reaching a point which would be harmful or damaging to the enclosed cables. A sensible guideline is to place heat barriers every 3 metres or between each junction of floor and ceiling.

Holes made in ceilings and floors prior to erection should be filled and sealed with a suitable material providing the same quality of fire resistance as the original. Fire barriers will prevent the spread of fire from one area to another.

Compliance with British Standards (Regulation 521–05–01)

Trunking must comply with BS 4678 or, alternatively, be made from an insulating material with a *Combustibility Factor* 'P' as recommended in British Standards 476 directive, Part 5. This publication can be obtained as a photo-copy from leading public lending libraries.

Exposure to water (Regulations 522–03–01 to 03)

When exposed to moisture, water or applied to an outside situation, steel trunking must be of the non-corrosive type and should never be placed in contact with other metals which are dissimilar. This will avoid any possible electrolytic action between the two surfaces. Where steel trunking is installed in such an environment, special precautions must be taken to prevent the ingress of water. In practice this usually requires a great deal of thought when selecting a trunking system to suit the environmental conditions.

Holes made to serve conduit arrangements should

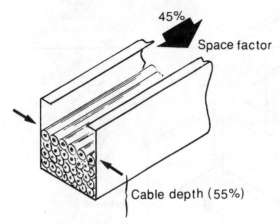

Figure 7.48. A space factor of 45 per cent.

TABLE 7.4 Once the factor has been calculated the size of trunking required may be determined

Unit factor (sometimes referred to as the Term)	Minimimum size of trunking required (mm)
738	25 × 75
767	37.5 × 50
993	25 × 100
1037	50 × 50
1146	37.5 × 75
1542	37.5 × 100
1555	50 × 75
2091	50 × 100
2371	75 × 75
3189	75 × 100
4252	100 × 100

always be made on the underside of the trunking and sealed around the periphery with a suitable sealer. As a practical precaution against accumulated condensation two or three 6 mm holes should be drilled in the trunking at its lowest level, (Figure 7.47) so that collected water may escape. Regulation 522–03–02 confirms.

How many cables? (Regulation 522–08)
Care must be taken to ensure that the number of cables laid or drawn into trunking during the period of installation or maintenance will not cause physical damage to either cable or trunking. Cables drawn into trunking via a right-angled bend could experience cable burn. This might go unnoticed until a fault condition is monitored during the testing period.

The *Wiring Regulations* will allow up to four 5 or 6 amp circuits to be bunched in trunking when protected by a semi-enclosed (rewirable) fuse or up to six circuits when protected by a high breaking capacity fuse (meeting the demands of BS 88), or a cartridge fuse to BS 1361. Alternatively, a miniature circuit breaker may be used.

When the installation is designed to accommodate more than the minimum number of circuits that may be bunched in trunking, a *correction factor* must be applied for grouping. Additional cables generate additional heat which could lead to serious damage to the installation if a correction factor is not applied. Applying a correction factor to the equation will effectively

increase the size of the conductors and therefore lessen the thermal effects.

Manufacturer's tables determining the size of trunking required for a given number of cables are readily obtainable. Alternatively, information may be gained from *Tables 5E* and *5F* of the *IEE's On Site Guide* which can be very useful, especially at the design stage of an installation.

How it works
Each size of conductor is allocated a number known as a *factor*. When all factors are added together the resultant figure is compared with a three- or four-figured number known as a *unit factor* or *term* (Table 7.4). The size of trunking to be used will have a unit factor equal to, or larger than, the sum of the combined cable factors.

As a practical example, consider the following problem:

Determine the size of steel trunking that would be suitable to accommodate ten single-phase circuits, each having a design current (I_B) of 18 amps. The installation is to be wired using single PVC-insulated copper cables with each circuit protected by a BS 88 over-current device (I_n) rated at 20 amps.

Solution:
Minimum current-carrying capacity of cables is

$$I_{\min} = \frac{I_n}{C_g} \qquad [7.3]$$

where I_n is the nominal current setting of the over-current device, and

C_g is the correction factor applied for grouping. (Obtainable from *Table 6C* in the *IEE's On Site Guide*.)

Applying *Table 6C*, ten loaded circuits are found to have a grouping factor of 0.48.

Substituting for known values, from Expression [7.3]:

$$I_{min} = \frac{I_n}{C_g} = \frac{20}{0.48} = 41.66 \text{ amps}$$

Consulting the *IEE's On Site Guide, Table 6D1, Installation Method 3* (Cables enclosed in trunking or conduit), the minimum size cable that may be used is 10 mm^2.

Referring now to Table 7.5 of this book, a 10 mm^2 single core stranded PVC-insulated cable has a factor of 36.3.

By adding each factor for each circuit the total unit factor may be evaluated and matched to Table 7.4. Hence:

Ten circuits (20 conductors × 36.3) = 726

A unit factor of 726 matches closely with the unit factor of 738 recommended for 25 mm × 75 mm trunking and would be suitable for the proposed installation.

This is the minimum recommended size that may be used but due consideration must be given to any additional wiring which may be carried out in the future.

Space factor

A *space factor* of not less than 45 per cent should be allowed for in installations where trunking is used. In practice this means that trunking should never be over-accommodated with cables. When all circuits have been drawn in, they must not occupy more than 55 per cent of the cross-sectional area of trunking. This means that 45 per cent of the cross-sectional area will be empty, as illustrated in Figure 7.48.

The *Wiring Regulations* call for '... *adequate means of drawing cable in and out*' (Regulation 522–08–02 confirms).

TABLE 7.5   Cable factors for trunking

Conductor size and type (cross-sectional area in mm²)		Factor (sometimes referred to as the Term)
Stranded	1.5 mm^2	8.1
Stranded	2.5 mm^2	11.4
Stranded	4.0 mm^2	15.2
Stranded	6.0 mm^2	22.9
Stranded	10.0 mm^2	36.3
Solid	1.5 mm^2	7.1
Solid	2.5 mm^2	10.2

Supports for trunking (Regulation 522–08–01)

Trunking must be supported and fixed adequately in order to avoid mechanical stress and damage to the insulated sheath of the conductors, their terminations and to the installation in general. As a guide, *Table 4D* of the *IEE's On Site Guide* may be consulted to provide maximum distances between supports for both steel and PVCu trunking systems. In practice, common sense will usually dictate, when working under site conditions.

Fire barriers (Regulation 527–02)

Regulation 527–02–01 calls for all holes made prior to the erection of trunking, to be effectively sealed when such walls or floors are designated as fire barriers. Material used to make good must provide the same quality of fire resistance as the original to prevent the spread of fire. An internal fire-resistant barrier must also be provided whenever the cross-sectional area of the trunking is in excess of 710 mm² (Figure 7.49)

In practice, glass wool is packed into the trunking at the designated fire barrier. If left proud, the glass wool will compress when the lid is firmly

Figure 7.49. A trunking fire barrier. (Reproduced by kind permission of *Walsall Conduits Limited*.)

secured. This will provide a satisfactory barrier against the spread of heat providing it is compatible with both trunking and cabling.

Partitioned trunking (Regulation 528–01)

Steel trunking systems containing *low-voltage circuits* supplied directly from the mains, *telecommunication circuits* and fire *alarm circuits* (Category 1, 2 and 3 circuits respectively), must be segregated by electrically earthed partitions continuously sustained to any common outlet serving the trunking. When Category 1 and 2 circuits are impractical to segregate, Regulation 528–01–05 requires all conductors to be insulated to the highest voltage present. This will mean that telecommunication cables must be wired using a 600 volt grade cable. Regulation 528–01–01 confirms.

However, mixed Category 1 and 2 voltages can be incorporated within multi-core mineral-insulated cable, as Regulation 528–01–01 will also verify.

Category 3 circuits, which include fire detection and alarm systems and emergency lighting installations, *must be* partitioned from other categories of circuits and separated from each other in accordance with Regulation 528–01–04 and BS 6701.

Compartmental trunking is shown in Figure 7.50.

Trunking as a current protective conductor
(This aspect has been taken from Section 543 of the Wiring Regulations)

Electrical continuity must be protected against mechanical, chemical and electrochemical debasement and permit other protective conductors to be taken off at tap-off points, where, for example, busbar trunking is used as a rising main.

The cross-sectional area of the steel trunking must be at least equal to the resolution of the following expression (Regulation 543–01–03 confirms):

$$S = \sqrt{\frac{I^2 \times t}{k}} \ \text{mm}^2 \quad \text{(Expression [7.1])}$$

where S is the cross-sectional area of the steel trunking in mm^2
  I is the value of the fault current in amps
  t is the operating time in seconds of the disconnection device under fault conditions

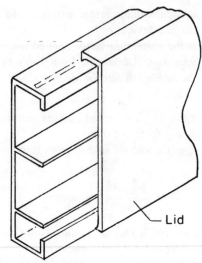

Figure 7.50. Compartmental trunking.

  k is a factor dependent on the material the trunking is made from; the initial and final temperatures and the insulation afforded. k factors may be taken from *Table 54E* of the *Wiring Regulations*.

Switching

Regulation 543–03–04 demands that no switching device is incorporated within the protective steel trunking. It must be continuous. Joints may only be incorporated within the trunking system for the purpose of testing. Steel trunking installations should never be used as a PEN conductor, i.e. Phase plus Earth and Neutral combined, as Regulation 543–02–10 will confirm.

Sockets in or served by steel trunking (Regulation 521–05–01).

Where a socket outlet is directly served by a steel trunking installation which also provides for a current protective conductor, the earthing terminal of the socket must be connected by an independent current protective conductor.

In practice, a green and yellow insulated flying lead attached to the earthing terminal serving the socket outlet is terminated at the back or side of the associated enclosure. Alternatively, it may be bonded to the main body of the trunking installation. This will then allow an earthing conductor to be present irrespective to whether the socket has been detached from the trunking or is securely in place as fitted.

Joints in steel trunking (Regulation 543–03)
When steel trunking forms part of, or is the sole provider of, the current protective conductor, joints within the system need not be accessible for inspection. It is therefore prudent to secure each coupling and trunk fitting thoroughly, and attach a copper bonding link to bridge each length and accessory installed. Protect from the possible effects of electrolytic action when installed in damp situations. Electrolytic contamination can be greatly reduced by applying a suitable coating before assembly or by using plated copper links.

Metallic cable tray installations

Cable tray is designed to take the place of trunking where steel-wired, armoured or mineral-insulated cables are grouped throughout a common route. It is widely employed for use in industrial or petrochemical installations where the use of heavy armoured cables is commonplace.

Figure 7.51 graphically illustrates a typical PVCu cable tray defined as a shallow semi-enclosure with not less than 30 per cent of a given surface area made up from perforated holes. Cables are attached by means of plastic cable ties, pierced galvanised heavy duty link-strap or plastic-covered aluminium strap.

Types
Today we are spoilt for choice and may select cable tray made from either mild steel, stainless steel, glass reinforced polyester, Lower Smoke Zero Halogen (LSZH) plastic or manufactured from PVCu. Finishes in steel cable tray include the following:

- expoxy painted
- chromate primed
- galvanised
- PVC coated
- red oxide

There are two categories of steel tray and choice is

Figure 7.51. PVCu cable tray. (Reproduced by kind permission of *Marshall Tufflex Limited*.)

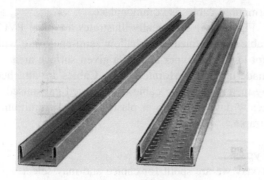

Figure 7.52. Heavy duty, return flange, cable tray. (Reproduced by kind permission of *Walsall Conduits Limited*.)

made according to the type of installation being carried out. Heavy duty tray is designed to accommodate larger or external installations in sensitive areas where a high degree of mechanical protection is sought. Figure 7.52 shows an example of this type of cable tray.

Light gauge tray can be used to accommodate mineral-insulated cables or small steel wire armoured control cables serving boiler houses or plant room installations.

Cable tray is generally produced in 2.5 and 3.0 metre lengths varying between 50 and 915 mm in width. When assembled it is secured by means of 6 mm diameter nuts and bolts anchored to custom-built mild steel brackets. Alternatively, steel channel supports in combination with stainless steel spring channel nuts can be used. This is often the more desirable method as it is far quicker than

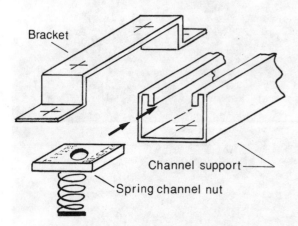

Figure 7.53. Channel supports or custom-built steel brackets can be used to provide support for cable tray.

making brackets from mild steel and painting them. Figure 7.53 shows in graphic detail the two methods which are generally available.

Brackets should be spaced between 1 and 3 metres from each other determined by the type of cable tray and the physical weight of the cables to be installed. If in doubt refer to the manufacturer for technical guidance, especially if PVCu tray is being used.

Working with cable tray

As with trunking, allowances must always be made for both route planning and decision making. Sufficient time should be given to mark out the proposed route and where, for example, right angles, riser bends, reducers and tee components are to be situated.

Long before work commences, the type and size of cable tray must be considered in order to avoid overcrowding. Ideally a 60 per cent space factor should be sought if the installation is covered or routed in a closed environment.

Width of cable tray

To obtain the required width of tray and, therefore, advanced suitable spacing between fixing brackets, the following expression may be used:

$$T_\phi = (d_1 + d_2 + d_3 + \ldots) \times (\% \text{ Space factor}) + d_t \qquad [7.4]$$

where T_ϕ is the required width of the tray in millimetres

$d_1, d_2, d_3 \ldots$ are the cross-sectional diameters of the cables

d_t is the sum of the overall total cross-sectional diameters.

As an example, consider the following:

A 3 metre section of cable tray of unknown width has to accommodate the following PVC steel wire armoured cables. Calculate the width of tray required, allowing for a space factor of 45 per cent.

1 in number, 16 mm^2 4-core cable
3 in number, 4 mm^2 3-core cable
7 in number, 1.5 mm^2 2-core cable

TABLE 7.6   Approximate cross-sectional diameters for PVC/
steel wire armoured cables

PVC/SWA/PVC	Type (overall diameters)		
Conductor c.s.a. (mm²)	Two core (mm)	Three core (mm)	Four core (mm)
1.5	11.7	12.3	13.0
2.5	13.1	13.6	14.5
4.0	15.1	15.8	17.7
6.0	16.5	18.1	19.2
10.0	20.1	21.2	22.8
16.0	21.4	23.1	26.3
35.0	24.5	26.9	30.1
50.0	27.5	30.1	34.6
70.0	30.0	34.2	38.4
95.0	34.8	38.5	43.4
120.0	37.2	41.5	48.1

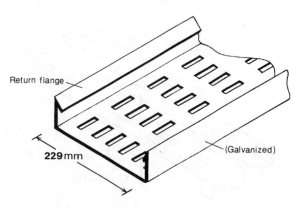

Figure 7.54. A section of heavy duty cable tray. See also Figure 7.52.

Referring to Table 7.6:

16 mm², 4-core cable has a mean diameter of

26.3 mm

4 mm², 3-core cable has a mean diameter of

15.8 mm

1.5 mm², 2-core cable has a mean diameter of

11.7 mm

Returning to Expression [7.4]:

$$T_\phi = (26.3 \times 1) + (15.8 \times 3) + (11.7 \times 7)$$

$$\times \, (0.45) + (155.6)$$

$$= (155.6 \times 0.45) + (155.6)$$

$$= 225.62 \text{ mm cable tray}$$

In practice, this would call for a 229 mm diameter heavy gauge tray. If installed outside and forming part of an agricultural installation, a hot-dipped galvanised tray supporting a return flange would be preferable (Figure 7.54).

Brackets and assembling techniques

Brackets should be sensibly positioned so that both tray and cabling are not exposed to undue mechanical stress. This is important in areas where cables are terminated as the total weight of the neighbouring installation could impose an unacceptable level of stress on both conductors and terminals.

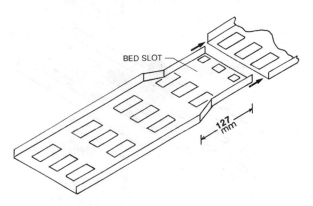

Figure 7.55. Cable tray bed slots.

With brackets firmly fixed, the tray can then be measured, offered up and secured by means of 6 mm galvanised nuts and bolts. Heavy duty tray is often accompanied with 'bed-slots' at one end of each 3 metre length (Figure 7.55) while other designs are reliant on 'fish plates' as a means of joining each section of tray together. Lightweight cable tray employs the 'lap over' method as described in Figure 7.56 and is best secured by means of four 6 mm galvanised nuts and bolts. Fixing data may be obtained from Appendix 4 of the *IEE's On Site Guide*.

Bonding

Copper bonding links must be firmly attached and bridged between each length of steel cable tray or accessory to provide adequate means of electrical continuity (Figure 7.57). As steel cable tray is a conductive part of equipment which, under fault

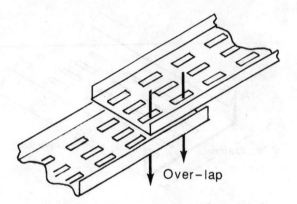

Figure 7.56. Lightweight steel cable tray is usually fitted together by means of the lap-over method.

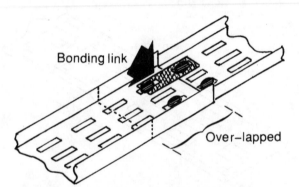

Figure 7.57. Copper bonding links must be fitted to the cable tray to meet the requirements of the *Wiring Regulations*.

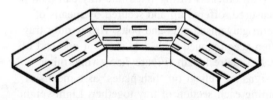

Figure 7.58. Cable tray angle (90°).

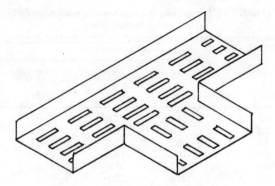

Figure 7.59. Equal tee piece.

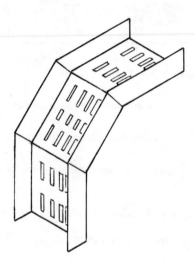

Figure 7.60. External riser bend.

conditions, could become 'live', it must be considered as an *exposed conductive part* and therefore be electrically bonded to earth to meet with the general requirements of the *Wiring Regulations*.

Cable tray accessories

Fortunately, there is now seldom need for site-assembled cable tray accessories, as an excellent choice of factory-made fittings are available. Good planning is the secret that will provide for the necessary component required. Listed below is a selection generally available. This will enable the task to be carried out far more efficiently than having to provide both time and material to build and assemble on site.

- Angle: 30°, 45° and 90° (Figure 7.58)
- Equal tee (Figure 7.59)
- Fishplate (for connecting sections of cable tray)
- External 90° riser bend (Figure 7.60)
- Flat bend: 30°, 45° and 60°
- Unequal tee

Site-built cable tray accessories

If site-built accessories are required, general

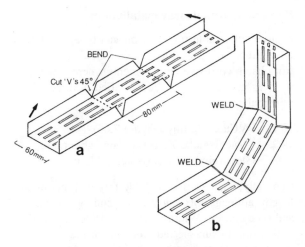

Figure 7.61. A site-built 90° inside riser.

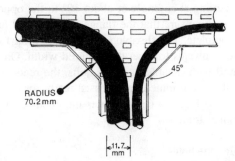

Figure 7.62. Choose cable tray junctions to accommodate the radius advised to serve the largest cable installed.

engineering principles must be applied. Figure 7.61 illustrates how a 90° internal bend can be constructed from a single section of heavy duty cable tray. Cutting two 45° 'V' slots in each flange will enable the cable tray to be internally bent to form a 90° right angle. Once formed, the prepared fitting should be either site welded or mechanically supported using flat iron and nuts and bolts to provide additional strength and stability. If the upright flange is welded, remember to apply a coat of cold galvanising to the joint. This will help to prevent corrosion occurring.

Equal tees

Cable tray junctions must be constructed to accommodate the radius advised to serve the largest cable installed. A factor of '6' must be applied to the overall diameter of the largest steel wire armoured cable to determine the minimum internal radius when a cable is formed into a right angle.

For example, a twin 1.5 mm^2 steel wire armoured cable has an overall cross-sectional diameter of 11.7 mm (Table 7.6). Multiplying by a factor of '6' will provide a minimum internal radius of 70.2 mm, as graphically illustrated in Figure 7.62.

Reducer

A simple reducer can be made by measuring the amount by which a cable tray is to be reduced and

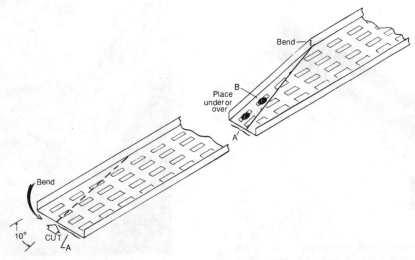

Figure 7.63. A simple site-built cable tray reducer. (A, proposed new width; B, bend the flap under the main body of the cable tray and secure.)

then cutting at an angle of about 10° to the opposite side, making sure that the flange is untouched and left complete. The flexible angled portion must then be bent inwards to meet the required width. Once secured with suitable nuts and bolts, the reducer may then be trimmed and dressed ready for assembly. Figure 7.63 illustrates this method in graphical form.

Cable exit holes

A small oblong hole cut into the main body of a cable tray is a simple way to provide an outlet for a cable leaving the system. Carried out correctly, this method will not mechanically weaken the tray to any extent.

Remove a measured portion of tray to accommodate the overall diameter of the cable to be routed out of the system and dress the sides with strips of 20 mm wide lead flashing or broad PVC edging strip. This will help to protect the cable from damage caused by vibration or movement due to expansion and contraction. Lead flashing may also be fitted to tray termination points providing adequate means of protection for cables leaving the tray system, as illustrated in Figure 7.64. Lead flashing is thin sheet lead; it is often supplied in small rolls and used extensively in the building industry.

Non-metallic cable tray installations

Non-metallic cable tray systems can be applied to most situations as well as meeting the demands from harsh corrosive environments where cable tray is necessary. These may be classified in three principal groups:

1. PVCu cable tray (Figure 7.65)
2. Lower Smoke Zero Halogen cable tray
3. Glass reinforced polyester cable tray

Generally, each section of cable tray is 3 metres in length, although this can vary depending on make and type chosen. Widths may be selected from 65 to 900 mm to suit the requirements of the installation. Each length of tray is assembled using factory-produced coupling units secured with stainless steel or PVCu nuts and bolts. Figure 7.66 pictures a selection of Lower Smoke Zero Halogen cable tray manufactured by the *Marshall Tufflex* company of Hastings, England.

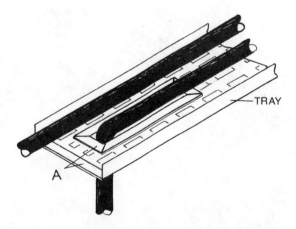

Figure 7.64. Remove a measured section of tray to suit the size of the cable to be re-routed. Dress the edging with lead flashing or PVC edging strip; A.

Figure 7.65. PVCu cable tray. (Reproduced by kind permission of *Marshall Tufflex Limited*.)

Figure 7.66. Lower Smoke Zero Halogen cable tray.

Expansion

A 1.0 to 2.0 mm gap should be allowed between each 3 metre length of non-metallic tray and washers should be fitted to each bolt securing both tray and couplings to fixing brackets. If this is overlooked, buckling could result when fluctuations in ambient temperature occur.

Cover

There are some systems which provide an optional heavy duty cover section which can be snapped into place by means of the flanges. This will not only protect sensitive cables but will help to increase the tray's overall rigidity. An example of tray cover may be seen in Figure 7.66. As with mini-trunking the cover is offered to the two edges of the flange and tapped into position using a rubber hammer. Using a metal hammer would probably cause damage to the lid or flange. To remove, prise off one end and peel back.

Advantages and disadvantages of using non-metallic cable tray

Advantages

The many advantages of using non-ferric cable tray outweigh the disadvantages. Listed are a selection of some of the benefits to be gained employing this type of system.

1. It is lightweight (density approximately 1.3 to 1.5).
2. It is a totally insulated system; bonding is not required.
3. It is easily worked.
4. A glass reinforced polyester cable tray is strong and may be used in high ambient temperature situations.
5. A Lower Smoke, Zero Halogen tray is ideally suited for fire alarm systems using LSZH-covered thermosetting cables.
6. There is a good selection of accessories and components.
7. All three categories of cable tray are virtually chemically resistant, but clarification should be sought from the manufacturer if the installation is thought to be at risk.
8. PVCu cable tray has a wide working temperature ranging from 253 to 315 K (−20 to +42 °C).
9. It will not corrode.
10. It can be fitted with an optional central cable separator for most makes of cable tray.
11. Once cut, all three types of tray can be easily dressed using glass paper or a fine file.
12. Internal cross-sectional areas range from approximately 1310 to 110 000 mm^2.
13. Lower Smoke Zero Halogen and glass reinforced polyester trays have an excellent working temperature range of between 233 and 405 K (−40 to +130 °C).
14. They are ideal for delicate or sensitive circuits as an optional heavy duty cover section may be fitted.

Disadvantages

1. PVCu cable tray is not suitable for installations where the ambient temperature is above 315 K (42 °C).
2. An installation might prove more expensive if a covered tray system is required.
3. A tungsten carbide tipped cross-cut saw is needed when Lower Smoke Zero Halogen or glass reinforced polyester cable trays are cut by hand. A suitable power cutting disc would be an alternative means of cutting.
4. Can become misshaped when exposed to

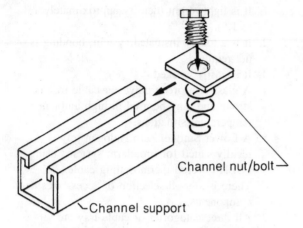

Figure 7.67. Glass reinforced polyester channel support will accommodate stainless steel spring channel-nuts to provide support for PVCu cable tray.

high ambient temperatures if allowance has not been made for expansion.

5. Poor chemical restance to:
   (a) ammonia liquid
   (b) nitric acid
   (c) potassium permanganate.
6. Guidance should be sought from the manufacturer if PVCu cable tray is to be used externally.

Brackets and cable supports

Fixing brackets supporting non-ferric cable tray are usually made from glass reinforced polyester, moulded into channel support. Cable tray is attached to the brackets by means of stainless steel spring channel nuts and 8 mm bolts as illustrated in Figure 7.67. Alternatively, steel site-built brackets could be used but it would be far more economical to use the recommended channelling.

Glass reinforced polyester is both corrosion resistant and fire retardant. The maximum recommended spacing for cable supports (clips, ratchet straps, etc.) may be taken from *Appendix 4* of the *IEE's On Site Guide*.

Summary

Cable trunking

1. Trunking profiles are manufactured to meet the demands of an installation (see Figure 7.2).
2. Miniature trunking is made from high-impact PVCu and is ideal for commercial and domestic applications (see Figure 7.6).
3. Spare sufficient time to reflect on the size of trunking required before marking out the proposed route. Mark where trunking accessories are to be installed.
4. Allow for thermal expansion.
5. Use high-impact PVCu trunking for light industrial and commercial installations.
6. Use steel trunking systems to provide a safe and sensible cable management network for industrial installations.
7. Never over-accommodate a trunking installation. Aim for a space factor of at least 45 per cent.
8. Vertically installed trunking should have steel retaining straps fitted to accommodate the cable.
9. Copper earth-links must physically bridge each length of steel trunking installed to provide adequate means of bonding the installation.
10. Avoid torsional and tensile stresses when installing cable in trunking as these could lead to problems in the future.
11. Fit fire barriers where required *after* the cables have been laid in the trunking.
12. Test the electrical resistance of the steel trunking before the installation is commissioned.
13. Typical busbar trunking is rated at 63 amps and is available for both single-and three-phase installations.
14. Site-built trunking accessories may be constructed whenever manufactured accessories are unavailable. Practice is the only way to gain a reasonably finished workpiece.
15. Use partitioned trunking for different category installations.

Cable tray

16. Select the type of cable tray to suit the application.
17. Cable tray is manufactured from either mild or stainless steel, glass reinforced polyester, Lower Smoke, Zero Halogen PVC or made from high-impact PVCu.

18. There are two main categories of cable tray: heavy duty and light gauge.
19. Fixing brackets can either be site built or supplied by the cable tray manufacturers.
20. Cable tray can be fitted with a central cable separator.
21. Allow for thermal expansion when installing.
22. Apply general engineering principles when site-built accessories are required.
23. Bonding links must be bridged between each length of steel cable tray or accessory to provide electrical continuity throughout the length of the installation.
24. Allow time for route planning before work commences.
25. Aim for a 60 per cent space factor if the installation is covered or routed in a closed environment.

Review questions

1. Describe briefly, 'cabinet casing'.
2. List two typical applications for mini-trunking.
3. How can thermal expansion be accommodated when installing non-metallic trunking?
4. What is the purpose of installing copper links to bridge each section of steel trunking or cable tray together?
5. How are fire barriers installed in trunking?
6. Why are busbar trunking components and accessories 'handed'?
7. Give a possible reason for high impedance recorded on a newly installed steel-trunking installation.
8. Where are PVCu skirting and dado busbar trunking systems often installed?
9. How can electrolytic contamination be controlled or eliminated?

10. Describe the term 'space factor'.
11. List three types of non-metallic cable trays which may be applied to meet the demands of an average installation?
12. Name five finishes in mild steel cable tray.
13. How may electrical continuity be ensured in steel cable tray systems?
14. Describe briefly the amount of gap which should be allowed to accommodate thermal expansion.
15. What type of cable tray installation would be ideal for sensitive circuits?

Handy hints

1. A simple but effective cable-tray saw can be easily made by shaping a suitable welding rod to accommodate a junior hacksaw blade. The saw can be used to form exit holes or to cut irregular patterns in the cable tray.
2. Steel cable tray or trunking which has suffered damage should always be made good and painted with an appropriate finish. This will help to prevent corrosion from occurring.
3. Bending springs used for heavy gauge PVCu conduit are colour coded green while those used for light gauge PVCu conduit are colour coded white. It is not practical to interchange.
4. Use only the correct equipment in high-risk flammable areas. Standard low-voltage (110 V) equipment and power tools will offer no protection against igniting flammable dust or vapour.
5. Seek confirmation in writing when variations or additional work is called for. Never rely on trust.
6. The *Wiring Regulations* have been adopted by British Standards and are listed as BS 7671.

8 Bathroom installations

In this chapter: Early times. Protection against electric shock. Protection against contact of live conductive parts. Wiring and the erection of equipment. Transformers serving shaver points and shaver lights. Bonding. Alternative wiring methods. Prohibited wiring methods.

Historical introduction

Earlier times

From the turn of the twentieth century to the period of the Second World War it was not uncommon to see a wall-mounted switch within reach from the bath, controlling an unshrouded pendant light and an occasional 5 amp unswitched socket outlet installed to serve an unguarded bathroom heating appliance.

Baths and basins were seldom bonded as it was considered that the plumbing installation was quite sufficient to provide adequate earthing arrangements.

The main earth conductor, unlike now, was connected to the service side of the stop cock, as illustrated in Figure 8.1, and this arrangement was considered to be satisfactory. Today, many of the existing metallic service mains have been replaced with polythene water piping, making it impossible to effect suitable earthing by means of the water company's service mains.

It is important that rooms containing a fixed bath or shower cubicle are electrically correct and have been wired with due regard to the *Wiring Regulations*. Rooms such as these can be some of the most potentially dangerous areas in a house and must therefore be treated accordingly.

Modern times
Today we are far more safety conscious and very much aware of the dangers of a cocktail of electricity, water and body resistance. Bonding is extremely important but often naively overlooked or not properly understood. Contact with moisture will dramatically reduce the resistance of the human body and produce lethal currents even when a lower voltage is present. The average resistance of the human body is approximately 6500 ohms, meaning that some will have much higher resistance, and others much lower. A bather saturated with water can reduce this figure by a staggering 75 per cent!

A combination of stray electrical currents and

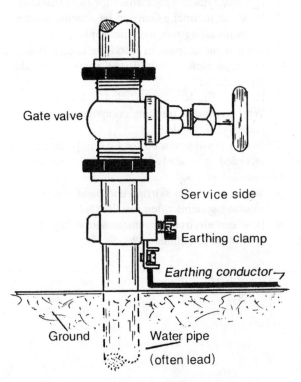

Figure 8.1. An old method of providing an earthing arrangement.

water can produce a deadly partnership. It is therefore wise to bond thoroughly as recommended by the *Wiring Regulations* in areas of high risk where water or condensation could be present. Bonding minimises the potential difference which, under fault conditions, would appear between simultaneously exposed conductive parts and accessible extraneous conductive parts. This concept will be examined in more detail later in the chapter.

Protection against electric shock

Equipment

Regulation 601–02–01 demands that no electrical equipment must be installed in the interior of a bath or shower basin. At first glance this might seem obvious but the regulation caters for just this sort of contingency.

SELV installations

Where *Separated Extra Low Voltage* (SELV) is used; for example, to serve a circuit of 12 volt recessed downlighters set into a bathroom ceiling or a wall-mounted extract fan, the low-voltage electrical mains equipment must be totally out of reach of a person using a bath or shower. Ideally, the transformer and low-voltage mains apparatus should be installed in an accessible ceiling void mounted on a suitable fire-resistant plinth (Figure

8.2). The transformer should be of a type which is completely insulated in a material that is capable of withstanding an AC test voltage of 500 volts (root mean square) for a period of one minute. If this is not possible, the transformer must be completely protected by means of barriers. If a fireproof insulated container is site built to meet the requirements of this regulation, thought should be given to ventilation as transformers can become very warm when used for long periods. (Regulation 601–03–01 confirms.)

Protection against indirect contact

Disconnection time

A fault condition to earth occurring in a room containing a fixed bath or shower must be capable of disconnecting the circuit within 0.4 second where electrical equipment can be touched simultaneously with other equipment or any extraneous conductive part. Examples are: hot or cold taps, an all-metal bath, water pipes and radiators. (Regulation 601–04–01 confirms.)

Bonding

Mechanical and electrical services installed in a room containing a fixed bath or shower cubicle must be both supplementarily and equipotentially bonded as recommended by Regulation 601–04–02. This is graphically illustrated as Figure 8.3.

The regulation calls for the following demands:

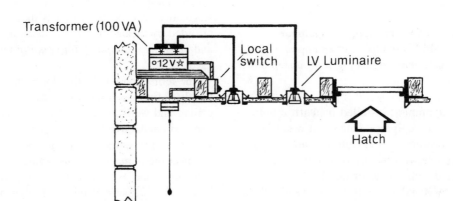

Figure 8.2. Ideally, mount the transformer and low-voltage mains apparatus in an accessible ceiling void.

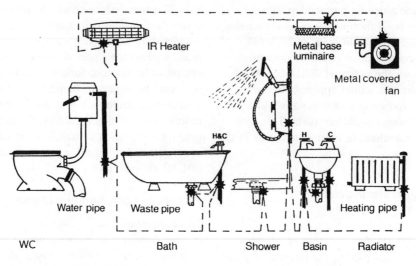

Figure 8.3. Bonding as recommended by Regulation 601-04-02.

1. Supplementary and equipotential bonding must be carried out between simultaneously accessible exposed conductive parts. (*Example: the steel body of an infrared wall heater and the metallic infrastructure of a shaver light.*)
2. Between exposed conductive parts and simultaneously accessible extraneous conductive parts. (*Example: the steel body of an infrared wall heater and a central heating radiator.*)
3. Between simultaneously accessible extraneous conductive parts. (*Example: a central heating radiator and a metal bath.*)

This regulation does not apply to equipment supplied by SELV circuits – for example, a 12 volt lighting arrangement originating from a step-down transformer.

Electrical equipment installed beneath a bath

When electrical equipment, such as a water pump, is installed beneath a fixed bath it should only be made accessible with the use of a tool. A bath panel secured with screws or turn buckles is demanded by Regulation 601–04–03; types which clip on are not acceptable when electrical apparatus is housed beneath the bath.

Protection against contact with live conductive parts

Direct contact

Regulation 601–05–01 demands that protection by means of obstacles or by placing an appliance completely out of reach must be used in a room containing a fixed bath or shower cubicle. Ideally, doubly insulated equipment should be used and positioned outside the area of risk. A practical example of this would be a *Class 2* wall-mounted fan heater sited out of reach of a person using a bath.

Indirect contact

Regulation 601–06–01 demands that protection afforded by means of a *non-conducting location* and *earth free equipotential bonding* must not be used.

The term 'non-conducting location' can be briefly defined as a room or area where floor, ceiling and walls are electrically insulated and all exposed conductive parts are spaced in a manner such as to prevent simultaneous touching by hands. To meet the demands of this regulation all extraneous and exposed conductive parts must not be less than 2 metres from each other or 1.25 metres where parts are out of reach. Although bonded to each other they must not in any way be

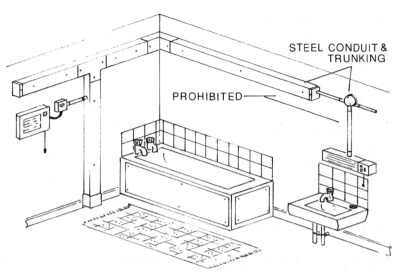

STEEL CONDUIT &
TRUNKING

PROHIBITED

Figure 8.4. Surface-mounted metallic conduit installations, trunking systems or bare copper mineral insulated cable should not be installed in a room containing a fixed bath.

bonded to an electrical earth. Clearly this type of protection is totally unsuitable for a bathroom installation.

Wiring and the erection of equipment

Regulation 601–07–01 forbids the use of the following in bathrooms:

1. Metallic conduit.
2. Steel trunking systems.
3. Exposed metallic cable, such as bare copper mineral-insulated cable.
4. Unsheathed steel wire armoured cable.
5. Non-insulated old-fashioned bonding cables.

It might be hard to imagine such a system installed in a domestic or commercial bathroom (Figure 8.4); however, the *Wiring Regulations* exist to protect and thus prevent such measures from being taken.

Control of electrical appliances

Regulation 601–08–01 requires all lighting switchings, dimmer units or other means of control to be inaccessible to a person using a fixed bath or shower. In a small bathroom a ceiling-mounted pull-switch could be used to satisfy the demands of this regulation. Alternatively, a wall-mounted wall switch may be sited outside the bathroom as illustrated in Figure 8.5.

Switched accessories serving a fixed heating appliance or towel rail must not be accessible to a person using a bath. In this type of situation a plain flex-outlet plate should be used, served by a suitable means of control outside the confines of the bathroom as illustrated in Figure 8.5.

Regulation 601–08–01 does not apply to the following situations:

1. Non-metallic switches may be installed within reach from the bath provided the source of energy is supplied from a SELV system not exceeding 12 volts AC or DC.
2. A shaver unit complying with British

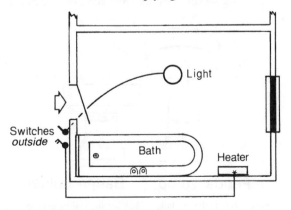

Light

Switches
outside

Bath

Heater

Figure 8.5. Switching arrangements serving a small room containing a fixed bath should be located outside the bathroom.

Standards directive 3535, where the earthing terminal is served with a protective conductor originating from the source of the supply. In practice, this will mean the distribution centre.

3. The nylon pull cord of a mains-operated pull-switch.
4. An insulated mechanical actuator mechanism linking a remotely operated switch.
5. Controls for a fixed instantaneous water heater complying to BS 3456, Section 3.9 (1979).

Transformer serving a shaver point or shaver light

Regulation 601–09–01 demands that no other means must be provided for the connection of an electric shaver other than a shaver point complying with BS 3535 of which the protective conductor serving the unit must originate from the source of the supply. This either means installing a dedicated circuit using PVC-insulated and sheathed cable or wiring a single bonding conductor originating from the distribution centre.

Socket outlets

A socket outlet forming part of a SELV system may be installed in a room containing a fixed shower cubicle or bath provided the following demands are met:

1. The supply does not exceed 12 volts AC or DC.
2. The socket does not have accessible metallic

Pendant drop   Batten holder

Figure 8.6. Bathroom lampholders serving ceiling pendants which are positioned within 2.5 metres from the bath must be shrouded in a moulded insulating material.

parts. (*A moulded non-standard socket would be ideal to meet the requirements of this directive.*)

Regulation 601–10–01 confirms that in no circumstances must a mains-operated socket be installed in a room containing a fixed bath. Any socket outlet installed in a room, other than a bathroom, which has a shower cubicle incorporated must be placed at least 2.5 metres from the shower cubicle to conform to Regulation 601–10–03.

Current-consuming equipment

Luminaires

Regulation 601–11–01 stipulates that lampholders serving ceiling pendants or batten holders which are positioned within 2.5 metres from a fixed bath or shower cubicle must be shrouded in a moulded insulating material as illustrated in Figure 8.6. Ideally a safety lampholder with a *Home Office Shield* to BS 5042, Part 1, should be used or a heat-resistant polyethylene terephthalate batten holder with a skirt may be fitted as an alternative. If options are available, a totally enclosed, insulated, heat-resistant ceiling fitting would be an excellent choice to meet the requirements of this regulation.

Heating appliances

Fixed heating appliances must be positioned so that a person using the bath or shower cannot touch the element. This regulation extends to the silica glass sheath shrouding the element of an infrared heater. Regulation 601–12–01 confirms. Figure 8.7 illustrates this demand in cartoon format.

Electric floor heating

Regulation 601–12–02 demands that when underfloor electric heating is installed in a bathroom or a room containing a fixed shower cubicle, a metal grid electrically connected to a local supplementary equipotential bonding conductor must shroud the heating elements.

Alternatively, a metallic sheathed heating element may be embedded in the floor; the protective sheath connected to a suitable supplementary bonding arrangement.

Figure 8.7. Fixed heating appliances should be installed so that a person using the bath cannot touch the element.

Bonding arrangements

Electrically bonding a bathroom installation is of paramount importance as without it lethal currents could cause death or injury to an unsuspecting victim. Contact with water lowers the body resistance; bonding minimises the electrical potential which under a fault condition could appear between conductive parts and will ensure that all metal work is of the same electrical potential. Zero volts!

Figure 8.8 illustrates the principal areas in which bonding should be carried out.

Main equipotential bonding conductor

Regulation 547–01–01 prohibits the use of insulated aluminium conductors for bonding purposes. Only green/yellow insulated copper cables may be used.

In installations other than those served with *Protective Multiple Earthing* (PME) arrangements, the size of the principal equipotential bonding conductor must not be less than half of the cross-sectional area of the supply authority's main earthing conductor serving the installation. If, for example, the main protective conductor from a TN earthing arrangement is 16 mm^2 (Figure 8.9), the main equipotential bonding conductor must be a minimum of 10 mm^2. On no account must the main equipotential bonding conductor be less than 6 mm^2 but it need be no greater than 25 mm^2 when calculated from the cross-sectional area of the supply authority's principal earthing conductor. Regulation 547–02–01 confirms.

Regulation 547–02–02 requires that the main equipotential bonding termination be made as near as practical to the gas and water service inlet points. These are the points where the services enter a building from the supply network buried in the road. In practice, this would mean that bonding would be effected within a distance of no less than

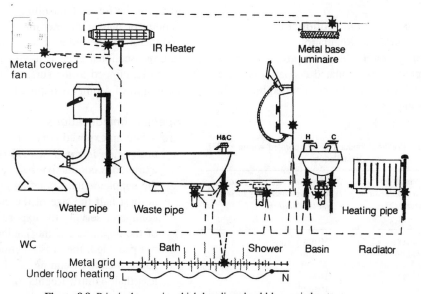

Figure 8.8. Principal areas in which bonding should be carried out.

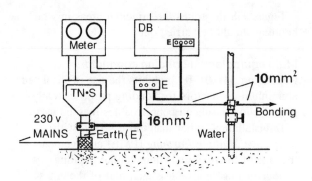

Figure 8.9. When the main protective conductor serving a TN earthing arrangement is 16 mm², the principal equipotential bonding conductor must be a minimum of 10 mm².

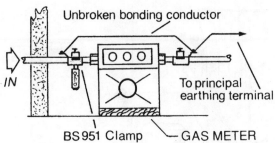

Figure 8.10. The *Wiring Regulations* require that the main equipotential bonding conductor be terminated as near to the gas and water service inlet points as practical.

600 mm from the stop cock and gas meter on the consumer's side of the installation, as illustrated in Figure 8.10.

Bonding a PME installation

Bonding arrangements serving a PME installation are governed by the size of the supply authority's main neutral conductor serving a consumer's switch gear. Reference is made to the *Wiring Regulations* and *Table 54H* and Regulation 547–02–01, whereas Table 8.1 provides details of the minimum size of bonding conductor required, taking into account the cross-sectional area of the supply authority's neutral conductor serving the customer's switch gear.

Supplementary bonding conductors

(Reference is made to Regulation 547–03.) Supplementary bonding must be provided by a reliable green/yellow insulated copper conductor which is supplementary to the main equipotential bonding arrangements.

TABLE 8.1 Bonding conductors incorporated within a PME installation

Size of supply authority's copper neutral conductor serving a customer's switch gear	Minimum size of the main bonding conductor to serve the installation
150 mm² and over	50 mm²
150 mm² to 95 mm²	35 mm²
95 mm² to 50 mm²	25 mm²
50 mm² to 35 mm²	16 mm²
35 mm² or less	10 mm²

Supplementary bonding conductors are designed to connect the following:

1. Two separate exposed conductive parts. (*Example: the tubular frame of an electric towel rail with the metal infrastructure of a shaver light.*)
2. Exposed conductive parts to extraneous conductive parts. (*Example: the tubular frame of an electric towel rail with the hot water pipes serving the central heating system.*)
3. Two or more extraneous conductive parts. (*Example: the hot and cold water supply with an all-metal bath.*)

In general terms, if mechanical protection can be given in the form of sheathing, sunken PVCu conduit or plastic channelling, the size of the conductor can be reduced to a minimum of 2.5 mm in cross-sectional area. If unprotected – for example, clipped to the surface of a wall – then a minimum of 4 mm² is required.

Special considerations

1. When an exposed conductive part (*the steel body of an infrared heater*) is connected to an extraneous conductive part (a cold water pipe), the supplementary bonding conductor must be at least half the size of the current protective conductor serving the exposed conductive part of the appliance. If mechanical protection is not provided, the cross-sectional area of the supplementary bonding conductor must not be less than 4 mm². Regulation 547–03–02 confirms; Figure 8.11 illustrates.

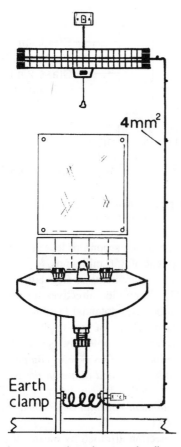

Figure 8.11. An unprotected supplementary bonding conductor must not be less than 4 mm² in cross-sectional area.

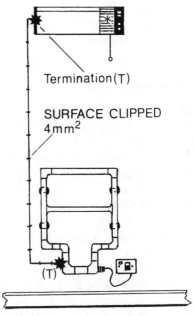

Figure 8.12. When two exposed conductive parts are bonded together the bonding conductor must be a minimum of half the size of the smallest current protective conductor serving either of the two exposed conductive parts.

2. When two exposed conductive parts such as an electric towel rail and a shaver light are bonded together, the bonding conductor must be a minimum of half the size of the smallest current protective conductor serving either of the exposed conductive parts of the two appliances. If mechanical protection is not provided, the cross-sectional area of the supplementary bonding conductor must be at least 4 mm². Regulation 547–03–01 con-firms. Figure 8.12 illustrates this regulation.

3. A supplementary bonding conductor connecting two extraneous conductive parts, such as a hot and cold water supply serving a bathroom basin, must have a cross-sectional area of at least 2.5 mm² if provided with mechanical protection. When protection is not supplied, the bonding conductor must not be less than

4.0 mm² in cross-sectional area (Figure 8.13). Regulation 547–03–03 confirms.

4. Regulation 547–03–05 advises that a flexible protective conductor incorporated within a short lead from a fixed accessory is deemed to provide a supplementary bonding connection to the exposed conductive parts of an appliance.

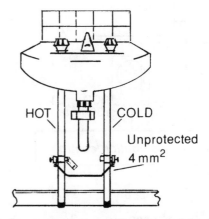

Figure 8.13. When two extraneous conductive parts are bonded together the bonding conductor must not be less than 4 mm² in cross-sectional area if mechanical protection is not provided.

As an example, consider a flex-outlet accessory serving a bathroom towel rail with a short length of heat-resistant flexible cable.

Alternative wiring methods

To reduce the risk of electric shock in a potentially hazardous area such as a bathroom, a local combined miniature circuit breaker/30 mA residual current device distribution centre can be used to protect all but the principal lighting circuit. This, ideally, would originate from a SELV arrangement served from a transformer sensibly located within an accessible ceiling void or suitable cupboard outside the area of risk. Figure 8.14 will help to explain this concept.

Regulation 412–06–01 advises that a residual current device (RCD) must not be relied upon as the sole means of protection should a fault condition to earth occur (Figure 8.15). Other measures must be incorporated within the installation to meet the requirements of this regulation. In summary these are:

1. The RCD must be of a type which provides a rated residual operating current not exceeding 30 milliamps and an operating

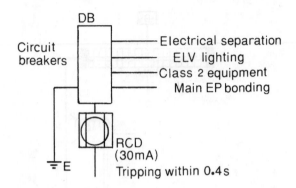

Figure 8.15. A residual current device must not be relied upon as the sole means of protection.

time of 0.4 second or less at a residual current of 150 milliamps.

2. The requirements of Regulations 418–02 and 471–08 provide directives for protection by earthed equipotential bonding and automatic disconnection of the supply voltage.

3. Protection afforded by use of *Class 2* equipment. (*This type of equipment is doubly insulated although the outer skin of such equipment can be metallic. Class 2 equipment must not be earthed, Figure 8.16.*)

4. Protection afforded by means of electrical separation as described in Regulations 413–06 and 471–12. To comply with this demand, the following measures must be taken:

    (a) The supply must originate from an isolating transformer complying with BS 3535. There must be no

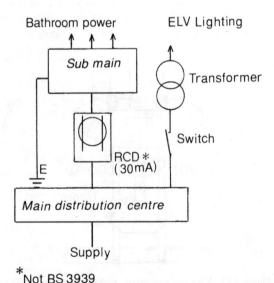

*Not BS 3939

Figure 8.14. An alternative wiring method for a room containing a fixed bath or shower.

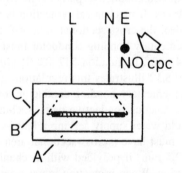

Figure 8.16. Protection afforded by use of Class 2 equipment. (A, load; B, insulated inner skin; C, pressed steel enclosure.)

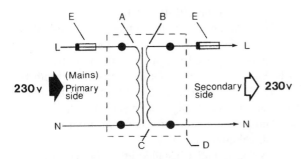

Figure 8.17. Protection afforded by means of electrical separation. (A, primary winding; B, secondary winding; C, isolating transformer; D, insulated enclosure; E, over-current protection.)

electromechanical connection between the primary and secondary windings or between the secondary winding and the body of the transformer and protective earthing circuit (Figure 8.17).

(b) Alternatively, a motor generator set which affords a similar degree of safety as the isolating transformer described above may be used.

(c) The voltage must not exceed 500 volts.

This is a very brief summary of the principal points required by this regulation. Although protection by electrical separation could be implemented it is not considered practical or realistic to serve a bathroom installation other than a BS 3052 shaver unit, and has only been mentioned as a theoretical alternative means of protection.

Prohibited protective wiring methods

Regulation 601–06–01 summarises the following protective measures which *should not be used* in locations containing a fixed bath or shower cubicle.

1. *Protection afforded by a non-conducting location as defined by Regulation 413–04.* This form of protection is unique, highly specialised and not recommended for day-to-day installation work. Briefly: the floor, ceiling and walls are electrically insulated and all exposed conductive parts are positioned in such a way as to prevent simultaneous touching by hands. All extraneous and exposed conductive parts

must not be less than 2 metres from each other or 1.25 metres where parts are unreachable.

Protective conductors are prohibited within a non-conducting location and socket outlets (*not applicable to bathrooms*) must be of a type which is without an integrated earthing contact. (*Regulation 413–04.*)

2. *Protection afforded by means of local equipotential bonding.* This is a highly specialised form of protection and should only be adopted in exceptional circumstances where technically specified and only under effective supervision.

   Briefly: all simultaneously accessible exposed extraneous conductive parts and exposed conductive parts are bonded together, *but not to an electrical earth*, as illustrated in Figure 8.18.

   Technical provisions must be made to ensure that upon leaving or entering the zone no one will be at risk from dangerous voltages, especially where a conducting floor, insulated from the electrical earth, is bonded to the earth free local equipotential bonding conductor. Inside the zone all conductive parts and extraneous conductive parts are the same potential and no danger can exist. Regulations 413–05 and 471–11 confirm these basic requirements.

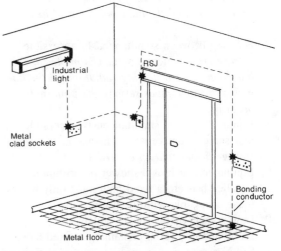

Figure 8.18. Protection afforded by means of local equipotential bonding.

These two protective wiring methods are prohibited in a room containing a fixed shower cubicle or bath, as required by Regulation 601–06–01.

Summary

1. Rooms containing a fixed bath or shower should be considered as a potentially high-risk area and therefore must be wired with due regard to the *Wiring Regulations*.
2. Exposure to moisture will reduce the resistance of the human body, and if direct or indirect contact is made with an electrical potential, lethal currents will be produced.
3. Bonding will minimise the potential difference under which fault conditions would appear between simultaneously conductive parts and accessible extraneous conductive parts.
4. No electrical equipment must be installed in the interior of a bath or shower basin. Low-voltage mains serving SELV arrangements must be completely out of reach from a person using a bath or shower.
5. An electrical fault to earth occurring in a room containing a fixed bath or shower must be capable of disconnecting the circuit within 0.4 of a second.
6. Bonding between exposed conductive and simultaneously accessible extraneous conductive parts. (*Example: metallic body of a shaver light and the bathroom central heating radiator.*)
7. Bonding between simultaneously accessible extraneous conductive parts. (*Example: the bathroom central heating radiator and the hot and cold water pipes serving the bath and basin.*)
8. Bonding between simultaneously accessible exposed conductive parts. (*Example: bathroom storage heater and an electric towel rail.*)
9. Equipment, such as a shower pump motor, installed beneath a fixed bath must only be made accessible with the use of a tool.
10. Low-voltage mains socket outlets are prohibited in a room containing a fixed bath. SELV accessories are only permitted if finished in a non-metallic material such as PVCu.

Regulation 601–07–01 forbids the installation of metallic conduit and trunking in a bathroom. Switched fused connection units must be placed out of reach from the bath.

11. Shaver or shaver-light units complying with BS 3535 may be installed in a bathroom, but the earthing terminal must be served with a protective conductor originating from the distribution centre.
12. A socket outlet installed in a room containing a fixed shower cubicle must be placed at least 2.5 metres from the shower unit.
13. Lampholders positioned within 2.5 metres from a fixed bath or shower cubicle must be shrouded. Ideally, a totally enclosed, insulated, heat-resistant ceiling fitting would be a better choice to meet these requirements.
14. Electric floor heating must be provided with a bonded grid placed above the thermoplastic heating elements. Alternatively a bonded metal sheath heating element may be embedded in the floor screed.
15. Aluminium insulated bonding conductors must not be used.
16. The main bonding conductor serving a bathroom suite must be at least half of the cross-sectional area of the main service earth when originating from a TN earthing arrangement.
17. Supplementary bonding conductors can be a minimum of $2.5\,mm^2$ if mechanically protected. If unprotected, a minimum of $4.0\,mm^2$ is required.
18. A bathroom suite can be additionally protected by the use of a combined miniature circuit breaker/residual current device distribution centre. The principal lighting circuit should not be served by the residual current device.
19. *Class 2 Equipment* (doubly insulated) must not be earthed.
20. Protection afforded by means of electrical separation is not considered practical.
21. Prohibited wiring methods:
    (a) Protection afforded by a non-conducting location
    (b) Protection afforded by means of local equipotential bonding.
22. Fixed heating appliances must be positioned so

that a person using a fixed bath or shower cannot touch the element.

23. The *Wiring Regulations* require that the main equipotential bonding termination is made as near as practical to the gas and water service inlet points which enter the building.

24. Never rely on the main water pipe for an effective earth as much of the existing network of metallic service mains has now been replaced with polythene water piping.

25. Bonding arrangements serving a protective multiple earthing system are governed by the size of the supply authority's main neutral conductor serving the consumer's switch gear. Reference is made to Table 8.1.

Review questions

1. Define the approximate resistance, in ohms, of the human body.
2. State briefly the purpose of electrical bonding.
3. In what circumstances is it permissible to install an insulated socket outlet in a room containing a fixed bath?
4. Where electrical equipment can be touched simultaneously with other equipment or any extraneous conductive part, disconnection time upon fault condition to earth must occur within
   (a) 5 seconds
   (b) 0.4 second
   (c) 1.0 second
   (d) 0.5 seconds
5. Confirm the following statements:
   (a) Exposed conductive parts serving SELV equipment (*the steel body of a 12 volt luminaire*) must not be bonded to earth. TRUE/FALSE
   (b) It is advisable to use aluminium insulated bonding conductors in a bathroom installation. TRUE/FALSE
   (c) In special cases, protection may be afforded by means of a non-conducting location in a room containing a fixed bath. TRUE/FALSE
   (d) Class 2 equipment, when installed in a bathroom or room containing a fixed shower cubicle, must be bonded to earth. TRUE/FALSE

6. If the principal protective conductor serving a TN earthing arrangement is 16 mm^2; what size should the main equipotential bonding conductor be?
7. How is the main equipotential bonding conductor size determined in a Protective Multiple Earthing installation?
8. List two examples of extraneous conductive parts.
9. A supply authority's neutral conductor serving a Protective Multiple Earthing system is found to be 150 mm^2. What is the minimum size of the principal equipotential conductor that must be installed?
10. State the minimum distance a socket outlet may be placed from a shower cubicle which has been installed within a bedroom.
11. Where, ideally, would it be wise to install low-voltage mains equipment associated with a SELV lighting arrangement?
12. In what circumstances is it permissible to install electrical equipment beneath a fixed bath?
13. What type of conduit system may be installed in a room containing a fixed bath?
14. Name the type of lampholder which must be used if positioned out of reach but within 2.5 metres from a fixed bath or shower cubicle.
15. What exactly is a Class 2 appliance?

Handy hints

1. Check out the internal wiring arrangements of a combination distribution centre served by two independently integrally fitted residual current devices. Manufacturers have been known to make mistakes.
2. Whenever work is carried out on a combination distribution centre, all *final circuit* neutral conductors must be recognised and placed in their respective neutral connection blocks. Wrongly placed conductors will cause imbalance within the residual current device causing the RCD to activate.
3. If in doubt always consult a competent colleague for advice.
4. Chromium-plated machine screws can be

converted to *brushed brass* by rubbing the head with coarse emery paper.

5. Bonding conductors are best terminated mechanically using a suitably sized pre-insulated ring type termination. Always use the correct compression tool in respect to the size of the termination to be crimped.

6. Reduce the length of nylon switch cord controlling a bathroom shower unit to shoulder height then trim the lighting switch cord to about waist height. This will avoid confusion; especially in the dark.

7. British Standards directive 2754 classifies mechanical protection against electric shock as follows:

- *Class 0 Equipment*: Appliance or equipment provided with basic insulation.

Exposed conductive parts are not earthed.

- *Class 1 Equipment*: Appliance or equipment provided with basic insulation and where all exposed conductive parts are earthed.

- *Class 2a Equipment*: Appliance or equipment provided with two layers of insulation of which one is the outer casing. This class of appliance is not earthed.

- *Class 2b Equipment*: Appliance or equipment provided with basic functional and supplementary layers of insulation to protect from live conductive parts. This type of appliance is not earthed.

- *Class 3 Equipment*: Appliance or equipment provided with basic insulation operating on a SELV system. This class of appliance is not earthed.

9 Towards National Vocational Qualifications in electrical installation

In this chapter: National Vocational Qualification, NVQ levels of competence. Methods of assessment. Written testing techniques. What is objective testing? Types preferred by awarding bodies. Self-assessment questions. Examples of objective questioning. Rational guesswork. Timing techniques and examination specifications. Answers to self-assessment questions.

National Vocational Qualifications

National Vocational Qualifications (NVQs), aimed to produce qualified electricians, were launched in September 1994.

An NVQ is divided into *Units* and each Unit is subdivided into *Elements*. Each Unit obtained is a step nearer to a full National Vocational Qualification at the chosen level. An NVQ gained will provide evidence of the level of competence of the examinee.

Candidates are assessed at a practical level by people with experience within the electrical installation industry and at a theoretical level by the awarding body. Marks obtained can be used towards a unit of competence.

NVQ levels

There are three NVQ levels of competence available to the electrical installation engineering student, but a candidate will only be classed as *skilled* when all three levels are obtained.

In summary these are:

- *Level 1* Fabricating and Fixing Electrical Cable Supports. (*This level comprises 6 Units, subdivided into a total of 15 Elements.*)

- *Level 2* Installing Electrical Systems and Equipment. (*Level 2 comprises 6 Units, subdivided into a total of 13 Elements.*)
- *Level 3* Installing and Commissioning Electrical Systems and Equipment. (*This level comprises 11 Units, subdivided into a total of 27 Elements.*).

Assessment

A variety of assessment methods are used by traditional awarding bodies, such as the *City and Guilds of London Institute*, for testing electrical installation engineering competence. Objective questioning technique is not the only method of assessment used. In summary, others include:

1. Practical assessment in the workplace or approved training centre.
2. Oral questioning.
3. Written testing.
4. Product examination assessment techniques.
5. Field Evidence Recording (FER) in the workplace.

Practical testing centres are found throughout Britain; your college of technology or training centre, or even your place of work, may be approved to carry out NVQ assessment requirements.

Written testing

Where considered suitable, parts of the City and Guilds 2360 syllabus not covered by NVQ standards can be assessed by an awarding body using *objective questioning techniques*, which are often considered by many to provide a fairer method of evaluation. Before 1968 the emphasis was placed on penmanship, the candidate's ability

TABLE 9.1   Basic installation practices

These are seen as an approximate percentage of a typical assessment paper if all syllabus topics were considered. In practice, however, six to nine topics are included each year as means to assess.

Syllabus reference	Topic	Approximate % of total syllabus
01	Study of the electrical industries	8.8
02	Associated core science	16.2
03	Safety practice and procedures	7.1
04	Moving loads	2.3
05	Procedures for work preparation and completion	8.8
06	PVC/PVC mineral-insulated cable installation	7.6
07	PVC/SWA/PVC cable terminations	3.5
08	Steel and PVCu conduit installations	10.8
09	Steel trunking installation	5.3
10	Cable tray installation	2.3
11	Principles of circuit installation	6.0
12	Earthing equipment installations	3.5
13	Inspection and testing of installations	9.6
14	Associated electronics technology	8.2

to express himself or herself fluently in writing. This proved to be an unfair means of assessment as many potentially good electrians failed their final examinations.

Objective testing

Advantages

1. As the correct answer has been pre-established, marking is completely objective.
2. Candidates are not required to express the answer using their own words. Ability, technical understanding and awareness of the subject can be immediately identified.
3. Virtually all syllabus topics and theoretical concepts can be included in one set of assessment papers (see Table 9.1).
4. The use of computer technology allows results obtained from examination papers to be published far more quickly.
5. Each section of the syllabus is covered by a fixed number of questions. The level of technical difficulty is arranged through a system of pre-testing students in their final year of study at selected colleges of technology.

6. Examination results obtained from year to year can be compared much more accurately under a system of objective testing.

General criticisms

1. Multi-choice question papers appear to be too easy.
2. Students will resort to rational guesswork techniques.
3. It is far better to express an answer using written English.
4. Candidates are often inexperienced in dealing with objective questions and this could lead to confusion.
5. An examinee could be confused with *distractor* and *key* options leading to an incorrect answer.
6. It is very difficult to set balanced multiple-choice questions.

What is objective testing?

An objective question consists of a sequence of listed written items or diagrams of which only *one* will represent the correct answer. This completely eliminates biased or subjective marking and will provide far better evaluation results. The candidates are not dependent on the mood or state of mind of the examiner and all students are treated equally.

Types preferred by the Institute

The City and Guilds of London Institute favours the four option multiple-choice system of questioning, of which the following types are preferred:

1. Matching block
2. True or false
3. Assertion/reason
4. Multiple response
5. Item groups
6. Multiple choice (This type of question can test the recall of factual knowledge or understanding of the subject. Alternatively, it may be used to test the application of practical or theoretical procedures.)

The analysis of a multiple-choice question

A multiple-choice question can be broken down as follows:

1. *Stem*. This is the name given to the question offered which may be either suggested or presented directly. It is followed by a list of four answers, known as *options*, of which only one is correct.
2. *Distractor*. This is the term given to the three incorrect options which at first glance will seem quite plausible. Distractors are based on common errors; statements which are only partly true; common misunderstandings or proclamations which are either too explicit or too narrow to address the question properly.
3. *Key*. This is the term given to the one correct option.

As an example, consider the following health and safety question:

Which one of the following treatments is medically correct for minor burns? (Stem)

(a) Petroleum jelly	(*Distractor*)
(b) Cooking fat or butter	(*Distractor*)
(c) Cold clean water	(*Key*)
(d) Bicarbonate of soda	(*Distractor*)

The answer or 'Key' to this question is (c) *cold clean water*.

Matching block

This type of objective multi-choice question is compiled from two lists of symbols, statements or mathematical expressions. The candidate is asked to correctly match one list with the other. As an example, consider the following question and Figure 9.1.

Which one of the following groups correctly matches items shown in List 1 with those shown in List 2?

Group (i)	List 1:	A	B	C	D	E
	List 2:	3	1	6	5	4
Group (ii)	List 1:	A	B	C	D	E
	List 2:	6	1	5	2	3
Group (iii)	List 1:	A	B	C	D	E
	List 2:	3	6	1	2	4
Group (iv)	List 1:	A	B	C	D	E
	List 2:	3	1	6	2	4

List 1
(From BS 3939)

A	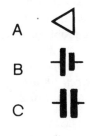
B	
C	
D	
E	

List 2

1. Capacitor
2. Bell
3. Telephone point
4. Switched socket
5. Wall-mounted luminaire
6. Battery

Figure 9.1. Matching block.

The key to this question is *Group (iii)*.

True or false

This type of question comprises a list of four statements or sentences. The candidate is asked to indicate whether or not they are technically correct by deleting either the word *true* or *false*. As an example, consider the following examination question:

Indicate whether or not the following statements are correct by deleting either the word 'true' or 'false' at the end of each statement.

1. Ceramic capacitors are generally used for high-frequency work such as radio, television and radar. TRUE/FALSE
2. Main gas and water services must be bonded within 900 mm from the consumer's side of the principal stop cock or gas valve. TRUE/FALSE
3. Wound rotor induction motors are often employed for use in large air-conditioning plants. TRUE/FALSE
4. A TN–S earthing arrangement is usually found in rural areas where the supply of electricity is delivered by a network of overhead cables. TRUE/FALSE

The key to this question follows:
1, *True*; 2, *False*; 3, *True*; and 4, *False*.

Multiple response

One or more of the choices provided is correct. The candidate is asked to judge which pair of items are true by annotating an entry in a secondary list. As an example, consider the following problem:

Which of the following elements or compounds shown in List 1 are found in a dry cell (a form of Leclanché cell). Options are provided in List 2.

List 1	List 2
A Ammonium chloride jelly	1. C and D
B Dilute sulphuric acid	2. B and D
C Solution of caustic potash	3. A and D
D Carbon and manganese oxide	4. B and C

The key to the problem is *A* and *D* (item 3).

Assertion/reason

This type of objective testing comprises a positive statement known as an *assertion* followed by a *reason*. The examinee must indicate whether the assertion and reason are both correct or not, and if correct whether the reason given is a factual interpretation of the assertion. As an example, consider the following question:

Assertion: Tungsten halogen lamps are often used for flood lighting but should not be handled with bare fingers.

Reason: Grease contamination can cause hot spots to occur which can reduce the working life of the lamp.

1. The assertion is false. The reason is true.
2. Both assertion and reason are true. The reason is a proper interpretation of the assertion.
3. The assertion is true. The reason is false.
4. Both assertion and reason are true. The reason is not a proper interpretation of the assertion.

The key to this question is *item 2*. Both the assertion and reason are correct and the reason is a proper interpretation of the assertion.

Item groups

This type of objective question has been designed

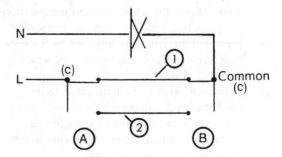

Figure 9.2. Item groups.

to test practical knowledge and experience gained during the candidate's period of training.

The question comprises a series of four items accompanied by a relative diagram, sketch or descriptive representation offered to the candidate. Marks are gained by annotating the correct item requested. As an example, consider the following and Figure 9.2:

1. *The type of switching arrangement shown in Figure 9.2 is a/an ...*
   (a) double pole switching arrangement
   (b) single-way switching arrangement
   (c) two-way switching arrangement
   (d) intermediate switching arrangement.
2. *The two conductors annotated as 1 and 2 are known as*
   (a) strappers
   (b) connectors
   (c) linkages
   (d) splicers.
3. *The BS 3939 graphical symbol served by the switching arrangement is*
   (a) an electric motor
   (b) an indicator lamp
   (c) a switched pendant light
   (d) a wall-mounted luminaire.
4. *The conductor linking the common terminal of Switch B with the right-hand side of the BS 3939 graphical symbol is known as the*
   (a) diversion conductor
   (b) switched wire
   (c) channel wire
   (d) change-over conductor.

The key to this question is: 1(c), 2(a), 3(d) and 4(b).

Multiple choice

Recall of factual knowledge

A question is offered with four options and the candidate must choose which option he or she considers to be correct. As an example, consider the following:

The maximum voltage that may be used to supply a SELV luminaire (BS-Class 3) is
  (a) 12 V AC or 25 V DC
  (b) 10 V AC or 12 V DC
  (c) 12 V AC or 12 V DC
  (d) 15 V AC or 15 V DC

The key to this question is *item (c)*: 12 V AC or 12 V DC.

Understanding of principles

This type of multiple-choice question is designed to test the candidate's understanding of electrical principles. A question is offered with four options of which only one is correct. As an example, consider the following:

Where a residual current device is incorporated within an installation supplied with a TN-CS earthing arrangement ...
  (a) a PEN (Phase–Earth–Neutral) cable may be used throughout the installation.
  (b) a PEN cable may not be used on the load side of the installation.
  (c) a PEN cable must only be connected from the load side of the residual current device to the distribution centre.
  (d) a PEN cable must only be used for circuits serving BS 2754, Class 2 equipment.

The key to this question is *item (b)*. A PEN cable may not be used on the load side of an installation where a residual current device is incorporated within an installation supplied with a TN–CS earthing arrangement.

Application

This type of multiple-choice question has been devised to test a candidate's ability to apply factual knowledge at a practical level. As an example, consider the following question:

Q1 *A residential driveway leading to a detached dwelling is to be illuminated with a 500 watt tungsten halogen luminaire. The most economical long-term method of control to serve the lighting arrangement would be ...*
  (a) a time switch with a spring reserve
  (b) a photo-electric cell
  (c) a passive infrared detector
  (d) an electronic time-lag switch.

(The answer to this question may be found at the end of the chapter.)

Examples of self-assessment questions

Most candidates find it useful and helpful to become familiar with the types of questions used in NVQ Unit Assessment papers. It is good practice to work through as many examples as possible. A candidate who is thoroughly acquainted with the design and arrangement of formal questions is less inclined to interpret them wrongly and will therefore avoid unnecessary incorrect answers. Familiarity enables the candidate to target the questions with confidence and helps to avoid confusion.

A few questions are repeated from year to year to provide in-house comparisons from results gained by candidates in previous examinations. However, the examples which follow are of the type which are favoured in assessment tests leading to a *Unit* or an NVQ award but they should not be regarded as representing the whole syllabus. The examples can also be used as *Self-Assessment Questions*. Answers are provided at the end of the chapter.

Example 1
What purpose does the outer sheath of a flexible PVC-insulated and sheathed cable serve?
  (a) a method to accelerate heat loss
  (b) to protect against harmful corrosion
  (c) to protect the cores from mechanical damage
  (d) to protect against electric shock

Example 2
Which simple solenoid, illustrated as Figure 9.3,

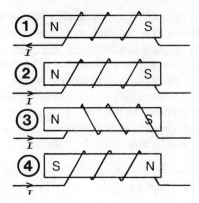

Figure 9.3. Illustration serving Example 2.

shows the correct polarity?
  (a) solenoid 2
  (b) solenoid 1
  (c) solenoid 4
  (d) solenoid 3

Example 3
The British Standard graphical symbol shown as Figure 9.4 is used in electronic engineering to represent
  (a) a zener diode
  (b) a thyristor
  (c) a transistor
  (d) a light-emitting diode

Figure 9.4. Illustration serving Example 3.

Example 4
In the year 1994, the supply voltage distributed by local UK electricity authorities was
  (a) 11 kV, 240 V and 110 V AC
  (b) 415 V, 240 V and 110 V AC
  (c) 132 kV, 415 V and 240 V AC
  (d) 11 kV, 415 V and 240 V AC

Example 5
Which one group of the following physical properties would be ideally suited for a general-purpose hacksaw blade?

	Hardness	Tensile strength	Brittleness
(a)	high	high	high
(b)	low	low	high
(c)	low	high	low
(d)	high	high	low

Example 6
The role of the capacitor shown in Figure 9.5 is to
  (a) prevent television and radio interference
  (b) provide energy to the starter switch
  (c) improve the power factor ratio
  (d) limit current on starting

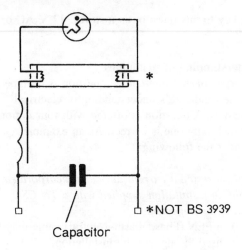

Figure 9.5. Illustration serving Example 6.

Example 7
Which one of the following expressions may be used to evaluate the rotor speed, in revolutions per second, of an AC induction motor?

(a) $\dfrac{\text{frequency}}{\text{number of pairs of poles}}$

(b) $\dfrac{\text{frequency}^2}{\text{number of pairs of poles}}$

(c) $\dfrac{\text{number of pairs of poles}}{\text{frequency}^2}$

(d) $\dfrac{\text{number of pairs of poles}}{\text{frequency}}$

Example 8

The moulded coloured covering serving mineral-insulated cable is to
- (a) provide means for circuit identification
- (b) prevent corrosion from occurring
- (c) provide protection from possible electric shock
- (d) provide protection from harmful ultraviolet rays

Example 9

How would the three monitoring indicator lamps shown in Figure 9.6 respond should the *Red Phase* (R) suffer a fault of negligible impedance to earth?
- (a) all lamps would be extinguished
- (b) all lamps would keep burning
- (c) lamp 'A' would burn but lamps 'B' and 'C' would be extinguished
- (d) lamp 'A' would be extinguished but lamps 'B' and 'C' would keep burning

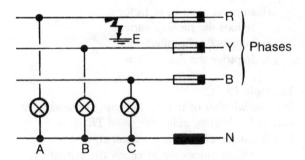

Figure 9.6. Illustration serving Example 9.

Example 10

Which one of the following four sets of colour sequences represents a value of 28 000 ohms when applied to a resistor with a tolerance rating of ±5 per cent
- (a) red, violet, yellow, silver
- (b) red, violet, blue, silver
- (c) red, grey, orange, gold
- (d) red, grey, red, gold

Example 11

When testing in accordance with the *Wiring Regulations*, the most suitable instrument to measure the resistance of each conductor serving a final ring circuit would be
- (a) an insulation tester set to 500 volts
- (b) an earth loop impedance tester
- (c) a low reading ohm-meter
- (d) a digital capacitance meter

Example 12

A newly repaired single-phase electric motor is found to vibrate when mechanically connected to a water pump. The primary cause of the vibration could be attributed to
- (a) an unbalanced armature
- (b) misalignment between motor and pump
- (c) motor incorrectly wired
- (d) an unbalanced pump rotor

Example 13

When the supply voltage is doubled serving a fixed value resistor, shown as Figure 9.7, current flow will be
- (a) doubled
- (b) halved
- (c) squared
- (d) unchanged

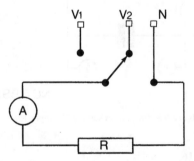

Figure 9.7. Illustration serving Example 13. (A, ammeter; R, resistor.)

Example 14

Three parallel fault conditions are measured between a common phase conductor and earth in a milking parlour.

Fault 1 was found to be 100 ohms
Fault 2 was found to be 500 ohms
Fault 3 was found to be 1000 ohms

The total resistance of the combined faults would be

(a) 1600 ohms
(b) 1000.92 ohms
(c) 400.92 ohms
(d) 76.92 ohms

Example 15

A low-pressure mercury vapour lamp circuit, illustrated as Figure 9.8, is connected to a low-voltage mains supply at terminals 1 and 2. What component must be connected to terminals 3 and 4 to allow the luminaire to function?
(a) starter switch
(b) radio interference suppressor
(c) a ballast unit
(d) a power factor improvement capacitor

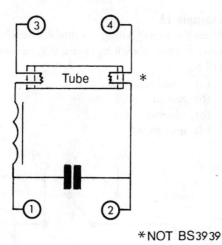

*NOT BS3939

Figure 9.8. Illustration serving Example 15.

Example 16

Figure 9.9 illustrates an arrangement of four

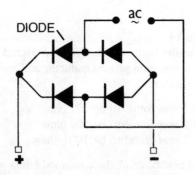

Figure 9.9. Illustration serving Example 16.

identical electronic components drawn to BS 3939 to represent a
(a) diode shunt in series formation
(b) full-wave rectifier
(c) parallel diode circuit
(d) half-wave rectifier

Example 17

It is essential to reduce the current rating of PVC-insulated and sheathed cable when exposed to high ambient temperatures because
(a) polarisation could occur
(b) the voltage drop could be unacceptable
(c) the insulation could break down
(d) the resistance of the conductor would decrease

Example 18

A capacitor can be connected across the main supply terminals serving a low-pressure mercury vapour lamp to
(a) improve the power factor
(b) reduce the running current
(c) suppress radio interference
(d) improve the light output

Example 19

Indicate whether or not the following statements are correct by deleting either the word TRUE or FALSE at the end of each statement.
1. DC machines are mechanically started by providing an inverse variable resistance to both armature and field coil in unison. TRUE/FALSE
2. The majority of motors manufactured today are designed to operate between ±15 per cent of the declared voltage. TRUE/FALSE
3. It is not practical to reverse the rotation of a shaded pole induction motor and it is often classed as non-reversible. TRUE/FALSE
4. Single-phase motor speed control is generally confined to split-phase capacitor-aided machines. TRUE/FALSE

Example 20

Which one of the following groups correctly matches the BS 3939 items shown in List 1 (Figure 9.10) with those shown in List 2.

List 1:   A   B   C   D   E   Group 1
List 2:   4   6   2   1   3

List 1:   A   B   C   D   E   Group 2
List 2:   4   6   2   1   5

List 1:   A   B   C   D   E   Group 3
List 2:   4   6   1   2   3

List 1:   A   B   C   D   E   Group 4
List 2:   4   5   2   1   3

List 1

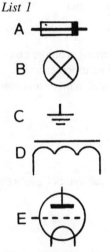

A
B
C
D
E

List 2

1. Inductor with core
2. Earth symbol
3. Triode valve
4. Fuse
5. Signal lamp
6. Motor

Figure 9.10. Illustration serving Example 9.20.

Example 21

When installed in a room containing a fixed bath or shower cubicle, which of the following items listed is not required to be bonded? Indicate by means of List 2.

List 1

A   Class 2 equipment
B   An all-steel bath with insulated
    legs
C   The hot water pipe
D   Exposed conductive parts of
    a SELV arrangement

List 2

1. B and C

2. A and D
3. B and D

4. A and B
5. C and D

Example 22

Assertion: Extra low voltage passive infrared detectors are often fitted in intruder alarm installations.

Reason: Their electronic circuitry will detect heat-movement through glass sandwiched between the detector and a transitive intruder.

1. The assertion is false but the reason is true.
2. Both assertion and reason are true and the reason is a proper interpretation of the assertion.
3. The assertion is true but the reason is false.
4. Both assertion and reason are true but the reason is not a true interpretation of the assertion.

Example 23

Figure 9.11 depicts a typical direct-on-line three-phase starter.

1. The three-phase supply is connected to terminals
   (a) 2, 4 and 6
   (b) 1, 3 and 5
   (c) 3, 5 and 7
   (d) 4, 6 and 8
2. The start button is annotated as letter
   (a) D
   (b) B
   (c) A
   (d) C
3. The load is connected to terminals
   (a) 2, 4 and 6
   (b) 4, 6 and 8
   (c) 1, 3 and 5
   (d) 3, 5 and 7

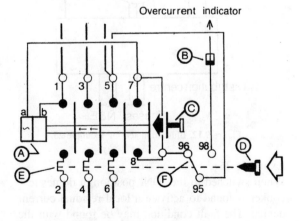

Figure 9.11. Illustration serving Example 23.

4. The thermal overload switch is annotated as
    (a) E
    (b) F
    (c) D
    (d) B

Example 24
Indicate whether or not the following statements are correct by deleting either the word TRUE or FALSE at the end of each statement.
1. Cables exposed to direct sunlight must be of a type that is resistant to damage produced by ultraviolet light. TRUE/FALSE
2. Flexible metallic conduit can be used as a protective conductor. TRUE/FALSE
3. A semiconductor, such as a diode, may be used as a means of isolation. TRUE/FALSE
4. Exposed conductive parts serving a SELV installation must be connected to earth. TRUE/FALSE

Example 25
In a well-balanced three-phase and neutral distribution centre, as shown in Figure 9.12, current flowing through the neutral conductor will be
    (a) the square root of the product of all phases
    (b) one-third of the total phase currents
    (c) no current flowing at all
    (d) one-sixth of the total phase currents

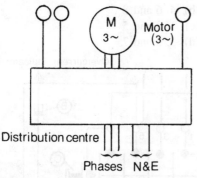

Figure 9.12. Illustration serving Example 25.

Example 26
When switched to the 'ON' position, a domestic cooker is found to activate a local residual current device. The fault condition may be found with the aid of
    (a) a milli-ohm meter
    (b) a continuity tester
    (c) removing one element at a time
    (d) an insulation tester

Example 27
To ensure that a supply has been completely isolated, a test should be carried out using
    (a) a proprietary voltage indicator
    (b) a suitably scaled ammeter
    (c) a wired lampholder and low-wattage lamp
    (d) a suitably scaled clamp meter

Example 28
Figure 9.13 illustrates how current, voltage and power may be measured by instrumentation.
1. The incomplete graphical symbol annotated as '1' represents
    (a) the load
    (b) the watt-meter
    (c) the volt-meter
    (d) the ammeter
2. The incomplete graphical symbol annotated as '2' represents
    (a) the ammeter
    (b) the load in series
    (c) the watt-meter
    (d) the volt-meter
3. The incomplete graphical symbol annotated as '3' represents
    (a) the ammeter
    (b) the volt-meter

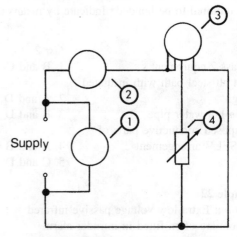

Figure 9.13. Illustration serving Example 28.

(c) the load

(d) the watt-meter

4. The incomplete graphical symbol annotated as '4' represents
    (a) the volt-meter
    (b) the watt-meter
    (c) the load
    (d) the ammeter

Example 29

List 1 itemises various ways in which electrical SI units may be evaluated. List 2 tabulates five random resolutions of List 1.

List 1	List 2
(a) $x = \dfrac{U}{R}$	1. Watts
	2. Amps
	3. Frequency
	4. Resistance
(b) $x = \dfrac{U^2}{R}$	5. Volts
(c) $x = \sqrt{(W \times R)}$	
(d) $x = \dfrac{U \times U}{W}$	

Which one of the following groups of expressions compiled from List 1 is synonymous with those compiled from List 2. This question is a matching exercise.

List 1: A B C D   Group 1
List 2: 2 1 3 4

List 1: A B C D   Group 2
List 2: 2 1 5 3

List 1: A B C D   Group 3
List 2: 2 1 5 4

List 1: A B C D   Group 4
List 2: 2 3 5 4

Example 30

Figure 9.14 illustrates the internal wiring arrangements, in schematic form, of a typical residual current device.

1. The graphical symbol labelled as 'F' represents the
    (a) tripping relay switch

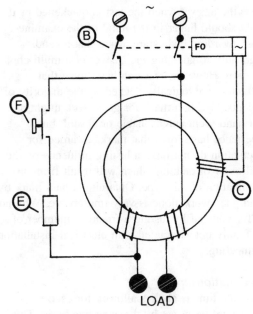

Figure 9.14. Illustration serving Example 30.

(b) test button facility
(c) over-current switch
(d) supplementary load switch

2. The graphical symbol labelled as 'B' represents the
    (a) test resistor
    (b) toroidal transformer
    (c) fault-sensing coil
    (d) tripping relay switch

3. The graphical symbol labelled as 'E' represents the
    (a) tripping relay
    (b) test resistor
    (c) test button facility
    (d) fault-sensing coil

4. The graphical symbol labelled as 'C' represents the
    (a) test resistance coil
    (b) auxiliary load coil
    (c) fault-sensing coil
    (d) over-current-sensing coil

Rational guesswork

One of the most common objections to objective multi-choice testing is that the slower student will guess the answer when the test is quickened or if doubt should linger in the mind of the examinee. Most awarding bodies are aware that a candidate might turn to guessing the answer of a multi-choice question. However, research has shown that distractors are logically deleted by the majority of students. Giving a mark for guesswork might at first hand seem unreasonable or absurd, but experience has shown that marks awarded for speculation do not have a lasting influence on the final order of ranking which will result from an examination of this type. Obtaining a pass mark by blind guesswork alone is very improbable. Why not try for yourself by testing a friend or member of the family not associated with electrical installation engineering.

Examination timing

Adequate time is always allowed for each assessment paper set by the awarding body. This removes the temptation to guess due to lack of time.

A typical two-hour assessment paper will contain between 60 and 70 multi-choice questions. These are not compulsory; however, the more questions answered the greater the candidate's chance to obtain a pass mark. It is wise not to spend any more than one and a half minutes answering each question. Time accumulated can then be used to review questions that have been passed over and to recheck options selected.

Answering the questions

It is not good practice to review both question and options at once. Good distractors are often plausible enough to seem correct and mistakes can be made when an option is chosen.

A wise approach is to read and thoroughly understand the question, keeping a brief mental answer on hold while reviewing the options. An option equivalent to an open-minded thought is far more likely to be correct than by choosing any other method.

Written papers

Examination papers requiring written answers usually contain from eight to ten questions and are generally of three-hour duration. Papers served with compulsory questions are often subdivided, each subdivision carrying a percentage of the total marks awarded to the question. All questions carry equal marks; however, individual marks awarded to each correctly answered section are usually shown after each sub-question.

Another type of assessment where written answers are required, comprises from eight to ten options from which six questions or more must be answered. This enables far greater freedom of choice and is popular among students. The time allowed for an examination paper of this type is usually three hours.

Timing a written paper

The period allotted for any type of written paper should be divided into equal parts, allowing an additional period in which checking may be worked through. As a guide, time assigned to each question may be calculated by use of the following expression.

$$\text{Allotted time} = \frac{\text{Duration of exam in minutes}}{\text{Number of questions} + 1}$$

[9.1]

As a practical example, consider the following:

A candidate sitting a C&G assessment paper is asked to answer ten compulsory questions over a three-hour period. Each question carries equal marks. Calculate the time that should be spent on each question and the time reserved for evaluation.

Referring to Expression [9.1] and adding known values:

$$\text{Allotted time} = \frac{3 \times 60}{10 + 1}$$

$$= \frac{180}{11}$$

$$= 16.36 \text{ minutes}$$

In practice, allowing 16 minutes to answer each question would provide additional reviewing time during the examination period.

Centre assessment

Practical assessment centres are scattered throughout the country; your own college of technology or training establishment could be approved to carry out NVQ assessment requirements.

Candidates are given a work sheet containing a number of practical assignments which are carried out at the centre where marking is also undertaken. Assignments are carefully structured to cover all knowledge evidence requirements of *Elements*, *Units* or NVQ's to be gained. As each *Element of Competence* is met it is ticked off by the official assessor and it is only when all criteria are satisfied that the candidate is said to have passed the practical assessment.

Both topics and assignments vary from year to year providing a changeable format so that no two examinations can be the same.

Answers to example questions

1. (c)	16. (b)	24. (2) False
2. (a)	17. (c)	24. (3) False
3. (a)	18. (a)	24. (4) False
4. (d)	19. (1) True	25. (c)
5. (d)	19. (2) False	26. (d)
6. (c)	19. (3) True	27. (a)
7. (a)	19. (4) True	28. (1c)
8. (b)	20. Group 4	28. (2a)
9. (d)	21. (2)	28. (3d)
10. (c)	22. (3)	28. (4c)
11. (c)	23. (1b)	29. Group 3
12. (b)	23. (2d)	30. (1b)
13. (a)	23. (3a)	30. (2d)
14. (d)	23. (4b)	30. (3b)
15. (a)	24. (1) True	30. (4c)

Answer to Q1: Passive infrared detector

Summary

1. A National Vocational Qualification is divided into *Units*, each unit is subdivided into *Elements*. A Level 1 or 2 NVQ both have 6 Units each while Level 3 comprises 11 Units.
2. Objective and written questioning is not the only method of assessment used. An awarding body will provide an approved assessment centre with a number of practical and theoretical assignments to be undertaken by NVQ candidates.
3. An objective question is formed from a sequence of listed items or diagrams of which only one is the correct answer.
4. Each question has three parts:
   (a) *Stem*. The name given to the question
   (b) *Distractor*. The name given to three incorrect options.
   (c) *Key*. The name given to the correct option.
5. Six types of objective questions could be included in a Unit assessment paper.
   (a) Matching block
   (b) True or false
   (c) Multiple response
   (d) Assertion/reason
   (e) Item groups
   (f) Multiple-choice
   Parts of the 2360 Syllabus not covered by the NVQ standards could be assessed by the awarding body using this type of objective questioning.
6. *A multi-choice question* is designed to recall factual knowledge, understanding or can be applied to test the application of practical or theoretical procedures.
7. *Item group questions* consist of a series of four items and a relative diagram, sketch or descriptive assertion offered to the examinee at the beginning. Marks are gained by annotating the correct item asked for.
8. *Assertion/reason questions* are constructed from a positive statement called an *assertion*, followed by a *reason*. The candidate must decide whether the assertion and reason are both correct or not, and if correct whether the reason given is a factual statement of the assertion.
9. *Multiple response questions* are arranged in two lists. One or more of the choices provided in the first list is correct. The student must judge which items are true by annotating an entry in the second list.
10. *True or false questions* comprise a list of four statements The candidate is asked to judge whether or not they are technically correct.

11. *Matching block questions* are compiled from two lists of symbols, statements or mathematical expressions. The examinee is asked to match one list correctly with the other.

12. It is good practice to work through as many self-assessment questions as possible.

13. A common objection to objective testing is that a slower student will resort to guessing the answer when the test is speeded up or if doubt arises in the mind of the candidate.

14. Research has shown that guesswork plays a minor role in providing answers for examination questions. Distractors are logically dismissed by the majority of students. Obtaining a pass mark by blind guesswork alone is very improbable.

15. A typical two-hour examination paper contains between 60 and 70 multi-choice items. Try not to review both question and option at once. Distractors are plausible enough to seem correct.
    - *Read the question*
    - *Keep a brief answer, mentally*
    - *Review the options*

    This type of paper usually incorporates questions from parts of the syllabus not covered by the NVQ standards and is set by the awarding body.

16. Papers served with compulsory questions not covered by NVQ standards are often subdivided for marking purposes.

17. Some assessment papers requiring written answers are provided with optional questions.

18. The time provided for a test requiring written answers should be divided into equal parts, allowing an additional period for checking and evaluation.

19. Each topic used in an assessment test is carefully chosen and represents a predetermined percentage of the total number of questions or assignments placed.

20. Some questions appear from year to year allowing in-house comparisons and evaluations to be made at a practical level.

21. Each Unit towards a National Vocational Qualification may be gained in a variety of ways. Not all are written or multi-choice testing methods. Level 2 NVQ knowledge evidence requirements which are directly related to the City and Guilds 2360 syllabus will be fashioned as assignments and carried out at authorised assessment centres and marked locally.

22. Virtually all Unit topics, whether practical or theoretical concepts, can be included in one set of assessment papers.

23. Computer technology permits results to be published more quickly.

24. Read and understand the questions thoroughly before attempting to answer.

25. NVQ assessment techniques include:
    - Written testing
    - Oral questioning
    - Product examination
    - Practical assessment
    - Field evidence recording in the workplace.

Review questions

1. Why is objective testing considered fairer?
2. Briefly describe the structure of an NVQ.
3. How many NVQ levels may be gained in Electrical Installation Engineering?
4. Specify three types of multiple-choice questioning that could be used to cover parts of the syllabus not correlated to the NVQ standards.
5. List three methods of assessment leading to a National Vocational Qualification award.
6. Why is it useful to become familar with the different types of questioning used in NVQ assessment tests?
7. Why are a few questions or assignments repeated from year to year?
8. Name three common objections to multi-choice questioning.
9. Name two types of testing centres approved to carry out NVQ assessment requirements.
10. There are three NVQ levels of competence available to the electrical installation engineering student. Name the title of Level 2.
11. Why is it not good practice to review both question and options at once in a multi-choice assessment paper?
12. Briefly describe a practical method of managing time allocated for an examination.

13. State briefly why sketches and diagrams should, when appropriate, accompany written answers.

14. How many Units must be gained to achieve a National Vocational Qualification Level 2 award?

15. What value can be placed on self-assessment questioning?

Handy hints

1. Set yourself a maximum time period for each assignment or question requiring a written answer but allow sufficient time for reviewing and, if necessary, correcting work done.

2. When assessment is carried out using multi-choice questioning techniques, memorise the answer before reviewing the list of options.

3. Distractors are designed to be plausible enough to seem correct and are based on common errors and misunderstandings.

4. Research has demonstrated that self-assessment questioning can help provide a sounder understanding in both practical and theoretical electrical installation engineering.

5. Prepare both physically and mentally for an assessment. Express yourself clearly to avoid ambiguity in written and oral examinations.

6. Assignments given to NVQ candidates at authorised assessment centres should be carried out in full in order to gain maximum marks. Written papers should be concise and to the point. Use simple sketches and diagrams to illustrate answers requiring penmanship. A picture is worth a thousand words.

   Mathematical short cuts should be avoided. Show all stages in your calculations.

7. Know your subject well!

Appendix A A brief history of achievements in electrical science

Name	Period	Achievement
Thales of Miletus	c. 600 BC	Discovered that by rubbing a stick of *amber* with dry fur he could attract flakes of straw to the amber.
Early Greeks	c. 402 BC	Concluded that all matter comprised minute, stable and indivisible particles. These they called atoms.
Pliny the Elder	AD 23–79	Discovered that powerful shocks would be received by anyone touching a torpedo fish.
Dr William Gilbert	1540–1603	Wrote a scientific paper entitled *De Magnete*, where he described in detail experimental work carried out with load stones and static electricity. He was first to coin the term 'electric' from the Greek word *elektron* for amber.
Otto von Guericke	1602–1691	Inventor of the first crude electromechanical machine. He discovered that electricity would flow from one conductor to another.
Dr Robert Boyle	1627–1691	Discovered that electrified objects, when placed into a vacuum, were still able to attract one another.
C.F. de C. du Fay	1699–1737	Concluded that glass was an excellent insulator and that objects charged with the same potential repelled each other while material charged with opposing polarities attracted each other. He was first to discover that the human body could conduct electricity.
Steven Gray	1696–1736	Discovered, in 1720, that electricity could be produced by the friction of silk, hair and wool. He was first to send a current along a thread approximately 293 metres in length by rubbing glass.
Sir William Watson	1715–1787	Developed the *leyden jar*. Discovered that electricity flowed along a wire in a way which seemed to be instantaneous.
Benjamin Franklin	1706–1790	Theorised that electricity was an invisible weightless fluid. He was able to prove that electricity and lightning were the same phenomenon.

John Canton	1718–1722	Discovered electrostatic induction.
Henry Cavendish	1731–1810	Calculated that the force of electrical attraction between two charged bodies varied inversely as the square of the distance between them.
C.A. Coulomb	1736–1806	Anticipated the fundamental law of current flow.
Luigi Galvani	1737–1799	Professor of anatomy at the University of Bologna, Luigi Galvani carried out work on 'animal electricity'.
Alessandro Volta	1745–1827	Developed the first acid battery.
André M. Ampère	1775–1836	Discovered new laws governing what was known as electrodynamics.
Hans Christian Oersted	1777–1851	Discovered that an electric current flowing in a wire adjacent to a compass would deflect the compass needle. He found that the way deflection occurred depended upon the direction of current flow in the wire.
Sir Humphry Davy	1778–1829	Demonstrated, in 1801, the principle of the arc lamp.
William Sturgeon	1783–1850	Produced the first electromagnet in 1825.
G.S. Ohm	1787–1854	Discovered that a current flowing in a circuit is proportional to the voltage applied and inversely proportional to the resistance at a constant temperature.
Michael Faraday	1791–1867	Discovered electrical principles governing the inductor and the dynamo.
Joseph Henry	1797–1878	Invented the principle of the telegraph but it was left to the American *S.F.B. Morse* (1791–1872) to develop the concept.
William Thomson (Lord Kelvin)	1824–1907	Developed theoretical electrical science and the knowledge required for laying submarine cables for wireless telegraphy.
James Clerk Maxwell	1831–1879	Pointed the way for wireless telegraphy and the telephone. Displayed a mathematical relationship between electricity and the speed of light.
Sir Joseph Thomson	1856–1940	Developed the electron theory. Measured the mass of subatomic particles and determined their polarity.
Heinrich Rudolph Hertz	1857–1894	Demonstrated that electromagnetic waves can be polarised. His work led the way to radio telegraphy.
Guglielmo Marconi	1874–1937	Developed the first practical telegraphic device using electromagnetic waves.
Albert Einstein	1879–1955	Explained photo-electric effects.

| Dr Ernst Meili | b.1913 | Developed the world's first ionisation type smoke detector for the *Cerberus Company* of Männedorf, Switzerland. |

Unfortunately space will only permit a small selection of famous names to whom we must all be indebted. Their dedicated pioneering work steered the way to unmasking many of the secrets of electricity. Their names will always be synonymous with electrical history.

Scientists such as *Dr Alexander Bell* (1847–1922), inventor of the telephone; *Nikola Tesla* (1857–1943), discoverer of electrical principles governing rotating magnets; the German physicist *Wilhelm Röntgen* credited for discovering *X-rays* and *Thomas Edison* who developed the first practical incandescent lamp, all have an important role to play in the brief history of electrical science. We owe them much.

Appendix B SI units

In 1960 the *General Conference of Weights and Measures* recommended a modern scientific system of weights and measures which would be internationally recognised. In due course the proposal was implemented and called *Système International d'Unités* (*International System of Units*), commonly abbreviated as SI throughout the world. This replaced the now defunct *Metre, Kilogram, Second* (*MKS*) system which was traditionally used before 1960.

SI units are divided into three categories:

1. Base units
2. Derived units
3. Supplementary units

Base units

Unit	Symbol	Quantifier
Amp	A	electric current
Candela	cd	luminous intensity
Kelvin	K	thermal temperature
Kilogram	kg	mass
Metre	m	length
Mole	mol	amount of substance
Second	s	time

SI symbols are always written in the singular form. For example, 1 mol and 100 mol, not 1 mol and 100 mols.

Derived units

Derived units are formed by combining two or more base units together. For example the unit of electrical charge, the *coulomb* (symbol C), is the product of the base units *current* and *time*. Not all derived units are provided with individual terms of reference. Some, for example, *magnetic field strength* (symbol A/m), are written combining both base units together (*amp per metre*).

There are many derived SI units which may be readily identified. Table B1 outlines a small selection of them.

Supplementary units

Just two units have been defined so far and both are dimensionless. The *steradian* (symbol *sr*) and the *radian* (symbol *rad*) are both supplementary units used for solid and plane angles respectively.

Multiplication factors and prefixes used in SI

Multiples and submultiples of units are offered in multiples of 10. Table B2 outlines the standard prefixes and multiplication factors used in association with SI units.

Practical applications

1   TW (terawatt) of power; that is,
    1 000 000 000 000 watts.
2   GV (gigavolt) of potential difference; that is,
    2 000 000 000 volts.
3   MΩ (megohm) of resistance; that is,
    3 000 000 ohms.

TABLE B1   Derived SI units

Unit	Symbol	Quantifier
Celsius	°C	Temperature
Farad	F	Capacitance
Hertz	Hz	Frequency
Henry	H	Inductance
Joule	J	Energy
Lumen	lm	Luminous flux
Lux	lx	Illumination
Newton	N	Force
Ohm	Ω	Resistance
Potential difference	V	Volt
Power	W	Watt

TABLE B2   Multiplication factors used in SI

Prefix	Definition	Symbol	Multiplication factor	Power
Tera	One million million	T	1 000 000 000 000	10^{12}
Giga	One thousand million	G	1 000 000 000	10^{9}
Mega	One million	M	1 000 000	10^{6}
Kilo	One thousand	k	1 000	10^{3}
Hecto	One hundred	h	100	10^{2}
Deca	Ten	da	10	10^{1}
Deci	One-tenth	d	0.1	10^{-1}
Centi	One-hundredth	c	0.01	10^{-2}
Milli	One-thousandth	m	0.001	10^{-3}
Micro	One-millionth	μ	0.000 001	10^{-6}
Nano	One-thousand-millionth	n	0.000 000 001	10^{-9}
Pico	One-million-millionth	p	0.000 000 000 001	10^{-12}
Femto	One-thousand-million-millionth	f	0.000 000 000 000 001	10^{-15}
Atto	One-million-million-millionth	a	0.000 000 000 000 000 001	10^{-18}

4   kJ (kilojoule) of energy; that is 4 000 joules.

5   cm (centimetre) of length; that is, 0.05 metre.

6   mA (milliamp) of current; that is, 0.006 amp.

7   μF (microfarad) of capacitance; that is, 0.000 007 farad.

8   ps (picosecond) of time; that is, 0.000 000 000 008 second.

Conversion factors

Other types of systems are very much in use today. Unfortunately we have not completely changed over to the metric system of measurement and at times we are faced with both imperial and SI units with which to evaluate a problem.

TABLE B3   Conversion factors

From imperial units	To SI units	Conversion factor
Fahrenheit	Kelvin	$(-32), (\times 5), (\div 9), (+273.15)$
Inches	Millimetres	$\times\ 25.4$
Inches	Centimetres	$\times\ 2.54$
Feet	Metres	$\times\ 0.304\ 8$
Yards	Metres	$\times\ 0.914\ 4$
Pounds (troy)	Kilograms	$\times\ 0.453\ 6$
Ounces (troy)	Grams	$\times\ 31.103\ 5$
British Thermal Units	Kilojoules	$\times\ 1.055$
Horsepower	Kilowatts	$\times\ 0.746$

(To convert degrees Celsius to Kelvin just add 273.15)

Table B3 presents a selection of yesterday's units which may be quickly converted into SI units with the aid of a pocket calculator.

Appendix C   Recommended tool kit

Basic requirements

Choose your tools with care, remembering that expensive tools can attract an opportunist thief. Never leave tools unattended and be wise when lending to site operatives. Try to insure against loss or theft but always read the conditions of insurance before committing yourself. Sometimes the listed exclusion clauses can be totally unsuitable.

Itemised alphabetically are the basic requirements to form an electrician's tool kit. Specialised tools, for example compression tools for copper terminations or holesaws and arbors, can be added as and when required.

1. Adjustable conduit grips
2. Adjustable spanner (medium)
3. Bolster chisel
4. Bush spanner (20 mm)
5. Chalk line string
6. Cold chisel (200 mm)
7. Combination metric square
8. Electric multi-speed percussion drill (110 volts)
9. File handle(s)
10. Files (25 mm flat and 8 mm round)
11. Floorboard chisel
12. Junior hacksaw and blades
13. Hacksaw and spare blades
14. Hammer (claw type)
15. Hammer (lump)
16. High-speed twist drills (assorted)
17. Insulated wire cutters
18. Mallet (plastic or rubber)
19. *Mole*® grips
20. Multimeter
21. Pad saw
22. Paint brush
23. Pencil
24. Philips screwdriver
25. Pliers (insulated)
26. Pliers (long nosed and insulated)
27. Plumb bob and line
28. *Rawlplug*® jumper (manual hole maker)
29. Screwdriver (terminal type)
30. Screwdriver (100 mm)
31. Screwdriver (150 mm)
32. Screwdriver (300 mm)
33. Spanner set (metric open ended)
34. Spanner set (metric box type)
35. Spanner set (small metric terminal type)
36. Spirit level (50 mm)
37. Spirit level (250 mm)
38. Suitable rule
39. Tap wrench (chuck type)
40. Tenon saw
41. Tool box (lockable)
42. Wood bits (10, 20, 25 and 32 mm)

Items such as stocks and dies, vices and heavy power tools, etc., are usually supplied by the employer.

Appendix D Answers to review questions

Chapter 1

1. In order to deal with large amounts of work, a company is divided into various departments. Each department is allotted certain responsibilities.
2. Joint Industry Board.
3. To manage and administer the relationship between employer and employee in the electrical contracting industry.
4. Technician, Approved Electrician, Electrician, Labourer.
5. Responsible to the director for the day-to-day running of all electrical installation contracts undertaken.
6. Stores list taken from the drawing or a computer printout of all material priced for the installation, supplied by the estimator.
7. Job holder and also in the care of a responsible operative.
8. Place a lap of coloured tape around the tool.
9. Variation orders, number of operatives on site, accidents, delivery of material, site visits and meetings, etc.
10. An unkempt appearance can create the wrong impression.
11. Vigilance, awareness of possible hazards and keeping to the rules and regulations.
12. Warning. Prohibition. Mandatory. Emergency/ Safe condition.
13. A white equal-sided vertical/horizontal cross placed on a square green background.
14. Can offer excellent career prospects, life assurance, payment upon accidental death, etc. Courses may also be attended. Provides an avenue for grievances.
15. Must be able to work without supervision and have a good working knowledge of current rules and regulations.

Chapter 2

1. Proton. Neutron. Electron.
2. Two.
3. Silver, copper, mercury, quartz.
4. (a) False. (b) False. (c) True.
5. The RMS value can be evaluated by dividing the square root of 2 into the maximum peak voltage.
6. Current limiter.
7. Step-up, step-down, and isolating transformer.
8. Copper losses. Iron losses.
9. A D-shaped copper bar, insulated by means of spacers, was drawn into a steel conduit and filled with bitumen.
10. Oil. Nuclear. Hydro. Gas.
11. Cost. Insulation considerations. Heating problems
12. (c).
13. Smaller conductors may be used. Current and heating effects are greatly reduced.
14. Electromagnetic radiation. Air molecule ionisation. Low-frequency sound waves.
15. Added to lightweight building blocks. Sold to manufacturers of cement products. Provides an infill for ground levelling schemes.

Chapter 3

1. (b).
2. It is difficult to monitor breaks which might occur during the life of the installation.
3. Mineral-insulated served with PVC covering or a suitable fire-resistant and flame-retardant Lower Smoke Zero Halogen cable such as Pirelli FP200.
4. (a) True. (b) False. (c) True. (d) False.
5. For fault monitoring
6. The tamper circuit is designed to protect the

wiring installation and detection devices from deliberate interference over a 24 hour period.

7. (a) False. (b) False. (c) True. (d) True.
8. Certain transmitted frequencies could trigger an alarm condition.
9. The panel is equipped with a small microswitch, fitted so that an alarm condition will be triggered when the door is forcibly opened. Often an additional microswitch is fitted to the rear of the control panel as a means of protection should the panel be prised from the wall.
10. Timed zone.
11. Category 3 circuit.
12. Mineral-insulated or a suitable fire-resistant and flame-retardant Lower Smoke Zero Halogen cable.
13. Acid baths. Conveyors. Rotating machinery. Turnstiles.
14. A heavier sized cable. Shorter cable runs. Fewer luminaires per circuit.
15 (d). Reference is made to Figure 3.43.

Chapter 4

1. Dial type. Digital type. Radio telemeter type.
2. To regulate heating loads.
3. Iron brass. Invar brass
4. Open types often accumulate grime, dust, humidity and atmospheric pollutants which prevent free movement of air over the active working parts.
5. Long unprotected runs of PVCu conduit will buckle under linear expansion due to high ambient temperatures.
6. The sum of the clockwise moments equals the sum of the anticlockwise moments.
7. A fixed point at which the resultant of all molecular weights act and where the force of gravity always passes.
8. The ratio of the length of arm (measured from the fulcrum to the effort) to the length of arm from the fulcrum to the load.
9. Totally dependent on the force of gravity. Weight can be measured by use of a spring balance.
10. Velocity ratio = $\dfrac{\text{Distance moved by effort}}{\text{Distance moved by load}}$

11. Stable. Neutral. Unstable.
12. (b).
13. (a) True. (b) False. (c) True. (d) False.
14. *Copper* and *iron* or *antimony* and *bismuth*.
15. Clinical. Spiral. Platinum resistance and Gas thermometer.

Chapter 5

1. Nitrogen. Argon.
2. Evaporated tungsten will deposit on the side of the bulb.
3. Up to 8000 hours.
4. False. (b) True. (c) True. (d) False.
5. Ultraviolet.
6. 1961.
7. Time switch, passive infrared detector and photo-cell.
8. Contactor controlled by a single lighting switch or a time-lag switch
9. Lumen.
10. Bayonet cap, small bayonet cap, Edison screw, giant Edison screw.
11. (c).
12. Used as a means of limiting current flow
13. The letter 'I' is printed on the body of a lamp served with an integral ignitor. Lamps requiring an external ignitor will have a letter 'E' printed on the side of the glass bulb.
14. A height of 2.75 metres from the ground.
15. It could cause nuisance tripping due to radio interference suppressors fitted within the control gear.

Chapter 6

1. Metals such as thoriated tungsten are able to release their surface electrons when heated to approximately 2273 K.
2. Temperature, area of cathode, material composition of the cathode, the ability of electrons to escape from the heated surface of the cathode and an air free envelope.
3. Germanium. Silicon.
4. p-Type material.
5. (b).
6. Carbon. Wire. Resistive film.
7. Age. Humidity. Temperature.

8. 6.6 ohms.
9. The dielectric.
10. (a) False. (b) True. (c) False. (d) True.
11. 60 F
12. 3 F
13. 30 H
14. As a means of amplification.
15. Visual indicator, display figures serving a clock and means of monitoring control equipment.

Chapter 7

1. Wooden trunking; used in the early 1900s.
2. Domestic and commercial applications.
3. Two supplementary elongated holes, one above the other and positioned 100 mm from the end of each length of trunking. Washers are placed behind each fixing screw. External expansion couplers may also be used.
4. Bonding continuity.
5. Suitable internal fire-resistant material such as glass wool is placed within the trunking where the system passes through fire-resistant structural material.
6. To ensure that correct polarity is maintained throughout the construction phase of the installation and that the component parts are assembled correctly.
7. Painted trunking with no bonding links fitted or a loosely fitted trunking coupler.
8. Used in domestic installations.
9. The maximum percentage of cross-sectional area permitted to be used by cables must not be greater than 55 per cent. This leaves a space factor of 45 per cent of unoccupied trunking.
10. By not bonding dissimilar metals together.
11. PVCu. Lower Smoke Zero Halogen. Glass reinforced polyester cable tray.
12. Chromate primed. Expoxy painted. Galvanised. PVC coated. Red oxide coated.
13. Bonding links
14. A 1.0 to 2.0 mm gap should be allowed between each 3 metre length of non-metallic cable tray.
15. Lower smoke, zero halogen or glass reinforced polyester cable tray as an optional heavy duty cover section may be fitted to provide additional protection.

Chapter 8

1. 6500 ohms
2. To minimise the potential difference under which a fault condition could appear between simultaneously conductive parts and accessible extraneous conductive parts.
3. When supplied by a 12 volt SELV system.
4. 0.4 second.
5. (a) True. (b) False. (c) False. (d) True.
6. $10 \, mm^2$.
7. Governed by the size of the supply authority's main neutral conductor serving a consumer's switchgear. (See Table 8.1.)
8. Bath. Radiator. Copper service pipes.
9. $50 \, mm^2$.
10. 2.5 metres.
11. In a ceiling void, mounted on a suitable fire-resistant plinth.
12. When the bath panel can only be removed with the aid of a tool.
13. PVCu conduit.
14. A safety lampholder with a *Home Office Shield* to BS 5042, Part 1, should be used or a heat-resistant batten holder with a skirt may be fitted as an alternative.
15. An appliance provided with two layers of insulation, one of which is the casing. This class of appliance does not require a protective conductor.

Chapter 9

1. As the correct answer has been pre-established, marking is therefore completely objective.
2. Levels 1, 2 and 3. Each Level is divided into *Units* and each Unit is subdivided into *Elements*.
3. Three.
4. Assertion/Reason. True/False. Item groups. Matching block.
5. In the workplace. Oral questioning. Written testing. Field Evidence Recording (FER) in the workplace.
6. Familiarity will enable the candidate to target the question with confidence and help to avoid confusion.
7. This system provides the awarding body with *in-house* comparisons from year to year.

8. Too easy. Students will resort to guesswork. Confusion could arise. It is better to express an answer in written English.

9. Colleges of technology. Training centres. The workplace.

10. *Installing Electrical Systems and Equipment.*

11. To avoid confusion it is wiser to establish the answer before reviewing the options.

12. Allotted time in minutes

$$= \frac{\text{Duration of test in minutes}}{\text{Number of questions} + 1}$$

13. A descriptive illustration is worth many words.

14. Six *Units*

15. Self-assessment questioning is an essential element in helping both learning and overall understanding in a chosen subject.

Index